PENTAGONS AND PENTAGRAMS

PENTAGONS

and

PENTAGRAMS

An Illustrated History

★

TEXT BY   **ILLUSTRATIONS BY**

ELI MAOR   **EUGEN JOST**

PRINCETON UNIVERSITY PRESS

Princeton & Oxford

Published by Princeton University Press
41 William Street, Princeton, New Jersey 08540
99 Banbury Road, Oxford OX2 6JX

press.princeton.edu

All Rights Reserved

Library of Congress Cataloging-in-Publication Data

Names: Maor, Eli, author. | Jost, Eugen, 1950– illustrator.
Title: Pentagons and pentagrams : an illustrated history /
    text by Eli Maor ; illustrations by Eugen Jost.
Description: Princeton : Princeton University Press, [2022] |
    Includes bibliographical references and index.
Identifiers: LCCN 2021041790 | ISBN 9780691201122 (hardback)
Subjects: LCSH: Pentagon. | Pentacles. | Mathematics—History. |
    BISAC: MATHEMATICS / Geometry / General | MATHEMATICS /
    History & Philosophy
Classification: LCC QA482 .M34 2022 | DDC 516/.154—dc23/eng/20211112
LC record available at https://lccn.loc.gov/2021041790

British Library Cataloging-in-Publication Data is available

Editorial: Hallie Stebbins, Kristen Hop, and Kiran Pandey
Production Editorial: Natalie Baan
Text and Jacket Design: Chris Ferrante
Production: Danielle Amatucci
Publicity: Kate Farquhar-Thomson and Matthew Taylor
Copyeditor: Jennifer McClain

Jacket Credits: *Front*: Eugen Jost, *Untitled*. *Back*: Eugen Jost, *The Golden Ratio*.

Frontispiece: Leonardo da Vinci, Dodecahedron

This book has been composed in New Century Schoolbook LT and Futura PT

Printed on acid-free paper. ∞

Printed in the United States of America

10 9 8 7 6 5 4 3 2 1

To my dear wife, Dalia, whose love and support
has meant so much to me over all these years

—ELI MAOR

To my grandchildren—Julian, Valentin, Laurin,
Carla, Elina, and Rafael—the joy of my old days

—EUGEN JOST

CONTENTS

Why the Pentagon?

Why the pentagon—no, not the defence
headquarters, but its eponym, the five-sided
polygon—continues to pester mathematicians.

—RICHARD MURCHIE[1]

ON SEPTEMBER 14, 1959, the Soviet spacecraft *Luna 2* crash-
landed on the Moon east of Mare Imbrium and became the
first human-made artifact to reach another world. Upon
impact, the spacecraft scattered around a cluster of small,
pentagonal-shaped metal pennants with the hammer and
sickle, the insignia of the Soviet Union (figure 0.1). The
pentagon—this ancient sign of brotherhood—thus became
humanity's greeting card on its first venture into the solar
system.

FIGURE 0.1. *Luna 2*

The story of the five-sided shape we call the *pentagon* spans at least 2,500 years. As far as we know, it was first studied by the Pythagoreans around the sixth century BCE. They were fascinated with it and saw it as a sign of good luck and fortune. According to tradition, they used the *pentagram*—a five-pointed star formed by tracing the pentagon's five diagonals with a single stroke of the pen—as an ID with which members of the sect greeted each other.[2] The shape has lost none of its mystical appeal over all these years: it is often used by Muslims, as can be seen in the exquisite glazed tiles, filled with pentagons and hexagons, that decorate many mosques throughout Central Asia, the Middle East, and North Africa (see plate 15). A pentagram appears on the national flag of Morocco (plate 8).

To many of us, the word *pentagon* means a regular five-sided polygon, in which all sides are of equal length and all angles of equal measure. In appearance, the regular pentagon, with its two lateral sides seeming to bulge outward, may not be as pleasing to the eye as a square, a hexagon, or an octagon. This is rather baffling, since the pentagon has ten symmetry elements (operations that leave it unchanged from its original orientation): five rotations, each through 72 degrees about the pentagon's center, and five reflections, each in a line running perpendicularly from a vertex to the opposite side. By contrast, a square has only eight symmetries (four 90-degree rotations, two reflections in the square's diagonals, and two reflections in the lines joining the midpoints of opposite sides). But appearances can be deceiving; as we shall see, the many internal relations among the pentagon and its five diagonals make this quite a challenging object to study.

But what the regular pentagon enjoys in symmetry, it lacks in another, equally important geometric feature: it cannot tile a plane without leaving gaps or causing overlaps. This is because two adjacent sides of a regular pentagon form an angle of 108 degrees between them; so for any number of them to join at a vertex, 108 would have to go an integral number of times into 360 degrees—which is not the case. A regular pentagon cannot tessellate the plane.

But what if we relax the restriction that all the pentagon's sides should have equal length and all its angles equal measure? It has been known for centuries that some non-regular convex pentagons can indeed tile the plane, and the systematic search for them makes for one of the most exciting stories in modern mathematics. As of today, fifteen types of convex pentagonal tilings are known to exist and are believed to exhaust all possibilities. The dramatic search for them, including the discovery of four hitherto unknown types by a woman with only one year of high school mathematical training, is told in chapter 7.

But let's return to the regular pentagon. If you join the pentagon's five vertices to its center (more precisely, the center of the circumscribing circle), you get a figure which, for lack of a better name, I'll call a *pentastar*; it is a common sight in nature, as evidenced by the numerous flowers or marine organisms with a fivefold rotational symmetry (see plates 1, 2, and 10). It is truly amazing how nature can create organisms with such perfect symmetry. Human-made designs with fivefold symmetries are not as common. The iconic pentagonal Chrysler emblem—once a familiar sight on the hoods of company-made cars—is now obsolete, but hubcaps with fivefold symmetry designs have recently become popular. In ancient Japan, costumes worn by members of the imperial court or dignitaries of social distinction were decorated with family crests, known in Japan as *kamon* or *mon*—decorative designs of floral and abstract geometric motifs, many with fivefold symmetry (figure 0.2). The tradition has since spread to wider circles outside the aristocracy and has survived to this day.[3] In the West, medieval fortresses were often built in a pentagonal

FIGURE 0.2. Japanese family crests

shape, with watchtowers at each of the five corners; we meet a rare surviving example in chapter 9. But the most famous pentagonal "fortress" of them all is a creation of our own time: *the* Pentagon in Arlington, Virginia, just outside Washington, DC, a landmark whose visual appearance and symbolic message are impossible to ignore.

The pentagon also has its place in the fine arts. Albrecht Dürer's famous engraving *Melencolia* (or *Melancholia*, 1514) shows a rhombohedron—a cube-like polyhedron with truncated corners, forming six pentagonal faces and two triangles (though the pentagons are nonregular; see figure 0.3). This enigmatic object is sometimes referred to as "Dürer's solid," but its allegorical meaning, as also that of

FIGURE 0.3. Albrecht Dürer, *Melencolia I*

several other geometric objects depicted in the drawing, is still debated today.[4] Stained glass windows are a common sight in numerous places of worship, but they almost always have eight-, twelve-, sixteen-, or twenty-four-fold symmetries. Plate 3 shows a rare exception, a pentagonal design at the Stephansmünster in Breisach, Germany—a Gothic-style cathedral originally built in the twelfth century as a Romanesque church. Closer to our own time, in Salvador Dali's famous painting *Sacrament of the Last Supper* (1955), the scene takes place inside what appears to be a dodecahedron, one of the five Platonic solids, having twelve faces, each a regular pentagon. We will have more to say about this work in chapter 3.

The year 1982 saw the discovery of a hitherto unknown kind of mineral, a quasiperiodic crystal with a tenfold symmetry (and, consequently, a fivefold symmetry as well). Up until then it was believed that the only possible symmetries in the inorganic world of crystals were two-, three-, four-, and sixfold symmetries, and nothing else. This finding earned its discoverer, Professor Dan Shechtman of the Technion—Israel Institute of Technology, the Nobel Prize in Chemistry—but only after a ten-year-long battle against the established scientific community, which had ruled that such a symmetry is impossible. His story—at once scientific and personal—is told in chapter 8.

★　⬟　★

Every morning, upon waking up, I stare at a five-bladed fan looking down on me from the ceiling. I imagine the five blades to be the radii of a circle in which a regular pentagon is inscribed, making me wonder how the ancient Greeks managed to draw this figure with only a straightedge and compass, in accordance with a tradition that Plato is said to have started around 400 BCE. Unlike an equilateral triangle, a square, or a hexagon, it is not at all obvious how to construct a regular pentagon. Indeed, almost nothing about the pentagon is as simple as meets the eye, from its geometric construction to the fairly involved calculations of the length of its diagonals or the area it encompasses. This must have greatly frustrated the early Greeks and may have started the aura of mystery about this shape, an aura that continues to this day. The secret to the construction is a number that has been variously called the *golden ratio*, the *divine proportion*, and *sectio aurea*. This number, about 1.618, is denoted by the Greek letter φ (phi) and is the subject of chapters 2 and 3.

Having dealt with geometric objects and patterns throughout our entire professional lives, we felt that the story of the pentagon deserved to be told in a broad historic and cultural context, with emphasis on its mathematical aspects. We chose to write it as an illustrated history, where pictures and visual form play an equal role to the written text. Our target audience is anyone interested in mathematics and its history, in nature and its influence on art and architecture, and in science in general. As with our previous book *Beautiful Geometry*, we limit the use of mathematics to an elementary level, "elementary" here meaning high school algebra and geometry. Eugen Jost has illustrated the book with color plates and black-and-white line drawings, while I have written the text. We hope the reader will enjoy our efforts.

We would like to thank the following persons for their help in writing and illustrating this book: Paul Canfield, Ron Lifshitz, Oded Lipschits, Ivars Peterson, Philip Poissant, Peter Raedschelders, Peter Renz, Kathy Rice, Moshe Rishpon, Doris Schattschneider, Daniel "daan" Strebe, Beth Thompson, Satomi Uffelmann Tokutome, and Douglas J. Wilson, as well as the staff of Princeton University Press for their continued support of our project. Thanks also go to the anonymous reviewers who read the manuscript and made many invaluable comments and suggestions. Last but not least, I want to thank my dear wife, Dalia—always by my side whenever I need some critical comments or ideas—for her constant support and encouragement in seeing this project come true. We truly appreciate their help!

A word about endnotes: When citing a source listed in the bibliography, only the title and author's name are given in the note. Otherwise, the full publishing information is cited.

Jerusalem, Israel, and Thun, Switzerland,
January 2022

NOTES AND SOURCES

1. Article published at https://the-gist.org/2016/10/why-the-pentagon-continues -to-pester-mathematicians/, 2016.
2. Sir Thomas Heath attributes this tradition to the comedy *The Clouds* by the playwright Aristophanes (ca. 446–386 BCE); see Euclid, *The Elements*, vol. II, p. 99.
3. See the articles "Kamon Symbols of Japan" at https://doyouknowjapan .com/symbols/ and *"Mon* (emblem)" at https://en.wikipedia.org/wiki/Mon _%28emblem%29. See also the book *Japanese Design Motifs: 4,260 Illustrations of Japanese Crests*, translated by Fumie Adachi, listed in the bibliography.
4. See the article "Melencolia I" at https://en.wikipedia.org/wiki/Melencolia_I.

PENTAGONS AND PENTAGRAMS

CHAPTER 1

Five

The Pythagoreans associated the number five
with marriage, because it is the sum of what
were to them the first even, female number 2,
and the first odd, male number 3.

—DAVID WELLS, *THE PENGUIN DICTIONARY OF*
***CURIOUS AND INTERESTING NUMBERS* (1986)**

EVEN AMONG THE FIRST TEN INTEGERS, five stands out. The
number one is, well, one, the generator of all integers.
Two is one doubled; it is the natural cycle that governs our
lives. We walk in steps of one-two, one-two, we breathe in
an inhale-exhale cycle, our daily activities are regulated by
the diurnal cycle of day and night, our body has a nearly
perfect left-right symmetry, and our sense of direction is
based on a left-right, forward-backward movement. The
Chinese yin-yang is a symbol of all things that come in
contrasting pairs—yes-no, on-off, good-evil, love-hate. Two
is the numeration base on which all our computers are
based, the binary system. We also note that two has some
unique mathematical properties: $2+2=2\times2=2^2$. And it
has the distinction of being the first prime number and the
only even prime. The exponent two is probably the most
common power in all of mathematics, appearing in the
Pythagorean theorem $a^2 + b^2 = c^2$, in the Mersenne numbers
$2^n - 1$ and Fermat numbers $2^{2^n} + 1$, and in numerous theorems in almost every branch of mathematics. It is just as
prevalent in physics as the exponent in all inverse-square

laws, and it stars in the most famous equation in all of science, $E = mc^2$.

Three is next in line, being the sum of one and two (although we often perceive it as a single unit in counting: 1-2-3, 1-2-3,…). Dances based on a triple meter are very common, from Haydn and Mozart's minuets to Beethoven's scherzos to the waltzes of the Strauss family. It is the first odd prime, as well as the first Mersenne prime ($3 = 2^2 - 1$) and the first Fermat prime ($3 = 2^{2^0} + 1$). It is also known as the "biblical value of π" due to a verse in I Kings 7:23: "And he made a molten sea, ten cubits from one brim to the other; it was round all about … and a line of thirty cubits did compass it round about." The "he" refers to King Solomon, and the "sea" alludes to a pond he ordered to be constructed at the outer entrance to Solomon's Temple in Jerusalem.

Next comes four, the smallest composite number and the only square integer of the form $p + 1$, where p is a prime (this is because $n^2 - 1 = (n + 1) \cdot (n - 1)$, a composite number except when $n = 2$). In the decimal numeration system, a number is divisible by four if and only if its last two-digit number is divisible by four (for example, 1536 is divisible by four because 36 is, but 1541 is not because 41 is not). The first of the regular or Platonic solids, the tetrahedron, has four vertices and four faces, each an equilateral triangle. Four colors are sufficient to color any planar map such that two regions sharing a common border will have different colors (this famous theorem was first conjectured in 1852 but was not proved until 1977). We view the world as comprising four dimensions, three of space (length, width, and height) and one of time, all merged into a single entity, spacetime. There are four cardinal directions, designated (in clockwise direction) as N, E, S, and W. Four is the number of letters in the ineffable YHWH (the so-called tetragrammaton), one of the names of God in the Judeo-Christian tradition.

Symphony Pathétique

Allegro con grazia   P. I. Tchaikovsky

FIGURE 1.1. The "limping waltz" in Tchaikovsky's *Pathétique*

AI HAI YO

CHINESE (MANDARIN) FOLK SONG

Āi   hāi yō,   āi   hāi yō,   āi   hāi yō hāi   yō   Wēn-hé tài-yáng zhào dà - dì,

bethsnotes.com

Xīn-nián yī - jīng__ dào,   Jiā jiā xìng-fú, Míng - nián__   hǎo__ shōu - chéng

FIGURE 1.2. Chinese folk song in pentatonic scale

We now arrive at five. It feels somewhat awkward to walk in steps of five, let alone to keep a five-beat rhythm in music. A quintuple meter of five quarter-notes per bar (denoted by 5/4) in classical music is quite rare; a notable exception is the "limping waltz" from Tchaikovsky's Symphony no. 6, *Pathétique* (figure 1.1). Similarly, a person accustomed to Western classical music may find it unnatural to listen to a piece played in a *pentatonic scale* of five notes to the octave. There are several versions of this scale; in one version, the notes are C, D, E, G, A, C′ (where C′ is one octave above C), comprising the intervals 1, 1, 1½, 1, 1½ (where 1 and ½ denote a full tone and a half tone, respectively); another version starts with C-sharp and follows the black keys of the piano, with the sequence of intervals 1, 1½, 1, 1, 1½. Pentatonic melodies can be found in much of African and Asian music. Figure 1.2 shows an example of a Chinese (Mandarin) folk song in a pentatonic scale.

But while it may feel awkward to count by fives in music, it actually comes quite naturally in daily life. This is due to the fact that we are born with five fingers on each hand. We are therefore endowed with a natural calculating device— literally, a "pocket calculator," considering that many of us like to hold our hands in our pockets on a brisk, cold day. And it doesn't need to be recharged, it never runs out of power, and it is always available and ready to be used. If this sounds a bit trite, consider that many cultures have developed a kind of "finger arithmetic," and all of us, at one time or another, have used our ten fingers to count or do some mental calculation. Indeed, the word *digit* literally means "finger"; so every time you use the adjective *digital*, remember that it comes from our built-in natural calculator.

The Romans had a special symbol for five: V, perhaps resembling a fully opened hand, while one, two, and three were written as I, II, III, obviously a visual image of the raised fingers representing these numbers. For quick tally-ing, the symbol 卌 is often used even today, as can be seen on many prison walls where inmates counted the number of days already served. For multiples of five, the Romans used the letters X = 10, L = 50, C = 100, D = 500, and M = 1,000. Other numbers were written in combinations of these symbols, such as IV (=4) and VI (=6). The fact that smaller values sometimes precede larger values but follow them in other cases made the Roman numeration system awfully difficult to compute with, but it has nevertheless survived well into the Middle Ages and beyond. Even today you can often see the groundbreaking date of a public building chis-eled in the cornerstone in Roman numerals. It was only in the Middle Ages that the Hindu-Arabic numeration system, with the numeral zero at its core, was gradually adopted in Europe and eventually became accepted internationally.

The Greek word for five is πεντε (spelled "pénte" in the Latin alphabet), from which numerous ancient and modern words derive; we encounter some of them later in this book. The Roman word for five was *quinque*, again the source of many ancient and current words. For example, *quincunx* describes a collection of five objects arranged in a square pattern, with one object located at the center and each of the others at a corner, as in the five-dot face on a die. And on the opposite side of the ancient world, the Chinese symbol for five was and still is 五 (pronounced like "me" in Mandarin), representing everything between heaven and earth and referring to the five elements that make up the universe: water, fire, earth, wood, and metal.

In the Hebrew alphabet, each letter is assigned a numerical value: א (aleph) = 1, ב (beith) = 2, ג (gimmel) = 3, ד (dalet) = 4, ה (heih) = 5, ו (vav) = 6, ז (zayin) = 7, ח (chet or het) = 8, ט (teth) = 9, and י (yod) = 10. Beyond ten, the system becomes additive (and read from right to left, as in all Semitic languages):

$$א" = 10 + 1 = 11, ב" = 10 + 2 = 12, ג" = 10 + 3 = 13,$$
$$ד" = 10 + 4 = 14.$$

But the next two numbers, 15 and 16, are written differently:

$$ט"ו = 9 + 6 = 15, ט"ז = 9 + 7 = 16,$$

this in order to avoid adjoining the letters י and ה, the first two Hebrew letters of the ineffable name YHWH, in accordance with the Third Commandment: "Thou shalt not take the Name of Hashem, your G'd, in vain." The remaining twelve letters after yod have the values 20, 30, 40,..., 100, 200, 300, 400.

The Hebrew word for five is חמש, pronounced "Ha'mesh."[1] Several words derive from it: חומש (Hu'mash), standing for the Torah—the Five Books of Moses, known in the Western world as the Pentateuch; חמישית (Hami'sheet, one-fifth);

מחומש (Mehu'mash, a five-sided polygon), and חמסה (Hamsah), an amulet resembling the open palm of a hand, symbolizing divine protection, fortune, and good luck; it usually comes in vibrant colors dominated by blue (plate 4), and is commonly found among Middle Eastern and North African cultures.

In the Talmud, the compilation of Jewish law written simultaneously in Jerusalem and in Babylon around the third century CE, it says "One should not donate more than a fifth of one's assets" (Babylonian Talmud, Tractate *Ketuvot*, p. 50a). The intention, no doubt, was to forewarn overgenerous donors against the possibility that they themselves might one day become dependent on charity.

★ ✿ ★

The ten fingers on our hands are the very reason why the decimal system has become the universal numeration system of the human race. Perhaps it isn't the best choice: had we been endowed with six fingers on each hand, a duodecimal (base 12) system would have been the natural choice, and a much better one indeed. For one, twelve has five proper divisors, 1, 2, 3, 4, and 6, whereas ten has only three, 1, 2, and 5. As a result, division in base 12 would be much simpler, avoiding, for example, a repeating decimal like 0.333 ... when dividing by 3. Second, many things in our lives already come in multiples of six—an egg carton contains twelve eggs, a pack of beer holds six cans, and our days and clocks are divided into twelve hours,[2] an hour has sixty minutes, and a minute has sixty seconds.

Around the middle of the twentieth century, the Duodecimal Society of America and a similarly named British society (both later renamed the Dozenal Societies) launched a vigorous campaign to change our numeration system from decimal to duodecimal. They issued decimal-to-duodecimal conversion tables, not just for integers but also for common and decimal fractions, special numbers like $\sqrt{2}$, π, and e, and even base 12 logarithmic tables. These were all well-intended goals, and logic stood on their side. In the end, however, five hundred years of familiarity with the decimal system have prevailed, and we are still holding on to the good old base 10 numerals.

Here is one small benefit of using base 10 as our numeration base. Because $2 \times 5 = 10$, we have $10/5 = 2$ and $10/2 = 5$. These last relations can be put to use for a quick, mental multiplication and division of a number by five: for multiplication, divide the number by two and move the decimal point one place to the right; for division, multiply the number by two and move the point one place to the left. For example, $38 \times 5 = (38/2) \times 10 = 19 \times 10 = 190$, and $47/5 = (47 \times 2)/10 = 94/10 = 9.4$. Yes, I know, everyone nowadays has a calculator on their smartphones, but still it is fun—and sometimes quicker—to do it mentally.

The ancient Babylonians used a hybrid of the base 10 numeration system for numbers from one to fifty-nine and a base 60 system—called the *sexagesimal* system—for numbers greater than or equal to sixty (presumably because sixty has ten proper divisors, 1, 2, 3, 4, 5, 6, 10, 12, 15, and 30, making division easier by reducing the need to use fractions). The Mayans preferred a smaller base: a hybrid system based on five for integers up to nineteen—the *quinary* system—and powers of twenty (the combined number of fingers and toes) for numbers greater than or equal to twenty (figure 1.3). The few written documents

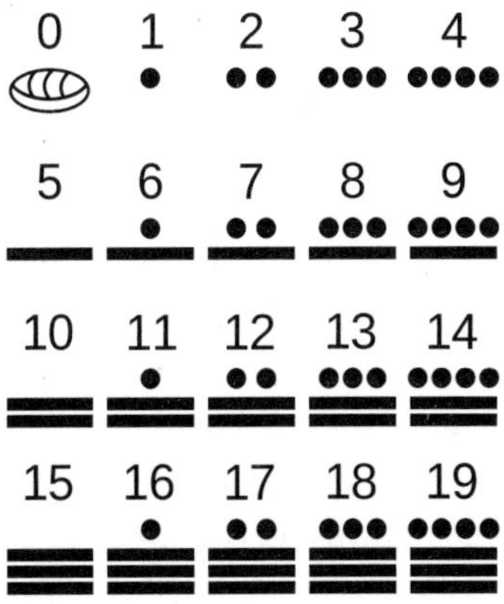

FIGURE 1.3. The integers one through nineteen in Mayan representation

that survived the Spanish conquest of their land include calendars and astronomical records using this *vigesimal* system.[3]

★ ⬠ ★

Before we turn to the number-theoretic properties of the number five relevant to the pentagon, here is a brief aside. The famous painting *I Saw the Figure 5 in Gold* by American artist Charles Demuth was first exhibited in New York in 1929 and is now in the permanent collection of the Metropolitan Museum of Art (see plate 5). Demuth (1883–1935) painted it as a tribute to a poem, *The Great Figure*, written by his friend William Carlos Williams describing a fire truck racing down the streets of New York on a rainy night. Demuth's painting became an American icon and appears on a US postage stamp issued in 2013. It also features on the cover of a mathematical novel, *Uncle Petros and Goldbach's Conjecture* by Greek author Apostolos Doxiadis, published in 1992. The title refers to German mathematician Christian Goldbach (1690–1764), who in 1742 wrote a letter to Leonhard Euler, then Europe's most famous mathematician, in which he claimed that every even integer greater than two can be written as a sum of two primes (sometimes in more than one way). For example, $4 = 2 + 2$, $6 = 3 + 3$, $8 = 3 + 5$, $10 = 3 + 7 = 5 + 5$, and so on. Euler, being occupied by more pressing problems, ignored Goldbach's letter; it was only found after his death in 1783. Despite its seeming simplicity and the fact that it has been confirmed for all even integers up to 4×10^{18}, the conjecture remains unproved. And while we are still on the artistic side of our story, Eugen Jost has depicted many of the daily occurrences of five in his painting *All Is Five*, which shows several whimsical allusions to various aspects of this number (plate 6).

★ ● ★

Five has some interesting mathematical features. It is the hypotenuse of the right triangle (3, 4, 5)—the smallest Pythagorean triangle and the only primitive one whose sides form an arithmetic progression (a primitive triple is one whose members have no common divisors other than one). Also, the sequence 5, 11, 17, 23, and 29 is the smallest sequence of five primes forming an arithmetic progression.

Five is the second Fermat prime ($5 = 2^{2^1} + 1$), and, consequently, a regular pentagon can be constructed with the Euclidean tools—a straightedge (an unmarked ruler) and compass. This is due to a discovery made by Carl Friedrich Gauss (1777–1855) when he was just nineteen years old: a regular polygon of n sides—a regular n-gon, for short—can be constructed with Euclidean tools if n is a product of nonnegative powers of 2 and/or *distinct* primes of the form $2^{2^k} + 1$, where k is a nonnegative integer. Primes of this form are called *Fermat primes*, after the great French number theorist Pierre Fermat (1601–1665).

The only regular polygons the Greeks knew how to construct with Euclidean tools were an equilateral triangle, a square, a pentagon, and a fifteen-sided gon, plus any polygons obtained from these by repeatedly doubling the number of sides (for example, the hexagon, octagon, and twelve-sided gon). Imagine the surprise when young Gauss added a new member to that list—a regular seventeen-sided polygon; that's because seventeen is the third Fermat prime ($17 = 2^{2^2} + 1$). As the story goes, Gauss was deeply impressed by this discovery and asked that a seventeen-sided gon be engraved on his tombstone after his death. But the stone cutter, fearing that a polygon with so many sides would be mistaken for a circle, chiseled a seventeen-pointed star instead. The original star is no longer visible, but Gauss's hometown of Brunswick, Germany, erected a statue in his honor, with a seventeen-sided star polygon engraved on its base. Plate 7, *Homage to Gauss*, is an artistic rendition of it by Eugen Jost.

Fermat conjectured that the expression $2^{2^k} + 1$ yields a prime for every nonnegative value of k. Indeed, for $k = 0$, 1, 2, 3, 4 we get the primes 3, 5, 17, 257, and 65,537, and therefore regular polygons with these numbers of sides are constructable with the Euclidean tools. Well, at least in principle. Even the seventeen-sided gon is fairly complicated to construct, and I wouldn't recommend anyone try the 257-sided gon.

Fermat's conjecture stood unchallenged until 1732, when Leonhard Euler showed that for $k = 5$ we get the Fermat number $2^{2^5} + 1 = 4,294,967,297 = 641 \times 6,700,417$—a composite number. As of this writing, it is not known if any other Fermat primes exist, leaving the possibility that there are other, as yet undiscovered regular polygons constructable with Euclidean tools. Needless to say, such polygons would have a huge number of sides, making any actual construction totally out of the question.[4]

Gauss's discovery provided a *sufficient* condition for constructing a regular n-gon with Euclidean tools. In 1837, Pierre Laurent Wantzel (1814–1848) proved that it is also a *necessary* condition, so the Fermat-prime polygons, and those obtained from them by repeatedly doubling the number of sides, are the *only* constructable n-gons. Thus, a fifteen-sided gon is constructable because $15 = 3 \times 5$, and both 3 and 5 are Fermat primes. But a seven-sided gon (a heptagon) is not, because 7 is not a Fermat prime. Nor is a fifty-sided gon, because $50 = 2 \times 5 \times 5$, and the double presence of 5 makes it ineligible. But a fifty-one-sided gon, practicably indistinguishable from its fifty-sided neighbor, *is* constructable, because $51 = 3 \times 17$, each of the factors being a Fermat prime.

Five is the fifth member of the Fibonacci series, a simple-looking sequence of numbers with many remarkable properties. The sequence starts with 1 and 1, then continues by adding the two previous numbers to get the next number:

$$1, 1, 2, 3, 5, 8, 13, 21, 34, 55, 89, 144,\ldots,$$

and in general

$$F_1 = F_2 = 1, F_{n+2} = F_n + F_{n+1}, n = 1, 2, 3,\ldots. \tag{1}$$

The sequence grows very fast: the twentieth member is 6,765, and the thirtieth member is 832,040. It is named after the Italian mathematician Leonardo of Pisa, born ca. 1170 to a Pisan merchant; he later became known by the name Fibonacci, meaning the son of Bonacci. In 1202 he published a book by the title *Liber Abaci* ("The Book of Calculation"), in which he advocated use of the Hindu-Arabic numeration system, known already for some time in the East but not yet widely accepted in Europe. The book became an instant hit and helped greatly in adopting the new system by merchants, then by scholars, and eventually by most of the learned world. The Fibonacci numbers appear in his book as a recreational problem: a pair of rabbits produce an offspring at the end of their first month and every month thereafter. The offsprings then repeat the same schedule. How many rabbits will there be at the end of the first year? It is easy to see that the number of rabbits follows the Fibonacci sequence, whose twelfth member is 144.[5]

It is somewhat ironic that Fibonacci's name is remembered today mainly for this little aside, rather than for his promotion of the Hindu-Arabic numeration system. His sequence enjoys numerous interesting properties, and a

FIGURE 1.4. The five Platonic solids, Bagno Steinfurt, Germany

scholarly publication, the *Fibonacci Journal*, is dedicated to their study. We will have much more to say about this sequence in chapter 3.

Five is the number of *Platonic* or *regular polyhedra*, symmetrical solids whose faces are all identical regular polygons that meet each other at the same angle (figure 1.4): the tetrahedron (four faces, each an equilateral triangle), the hexahedron, more commonly known as the cube (six faces, each a square), the octahedron (eight equilateral triangles), the dodecahedron (twelve regular pentagons), and the icosahedron (twenty equilateral triangles). That there exist exactly five regular solids—unlike the infinitely many regular polygons in the plane—is surprising and has made these solids an object of endless fascination (for a proof, see appendix C). The Pythagoreans were familiar with all five solids and knew how to construct them, using only the Euclidean tools. Four of these solids involve either equilateral triangles or squares, which are easy to construct; but the dodecahedron has pentagonal faces, whose construction is not at all obvious. It was this problem that most likely led them to discover the golden ratio or divine proportion—the key to constructing the regular pentagon.

NOTES AND SOURCES

1. It is somewhat difficult to transliterate the guttural Hebrew consonant ח (het) into English; it is variously written as "ch" or just "h."

2. Or twenty-four hours, known in the United States as "military time" but in common usage throughout the rest of the world, where no one has any trouble reading 17:00 as 5:00 p.m.

3. For more on the Mayan numeration system, see Georges Ifrah, *The Universal History of Numbers* (New York: John Wiley, 2000), pp. 44–46, 94–95, 308–12, 339; and Frank Swetz, *From Five Fingers to Infinity* (Chicago: Open Court, 1994), pp. 71–79.

4. Around 1980, when the first programmable calculators appeared on the market, I bought Texas Instruments' latest version, the SR 56 (the designation SR stood for "slide rule," until then the trade tool of every scientist and engineer for the past 350 years). It had a ten-digit display, so I programmed it to factor a number into its prime factors, punched in 4,294,967,297 and hit the "start" key. For the next 28 minutes the machine did its calculations, and then the smaller of the two factors, 641, appeared in the display, to my great delight. Needless to say, a modern computer can do it in a tiny fraction of a second. (There are several factorization sites available online, such as Prime Factors Decomposition at https://www.dcode.fr/prime-factors-decomposition.)

5. The sequence can be extended to negative indices as well, by rewriting equation (1) as $F_n = F_{n+2} - F_{n+1}$: ..., $-8, 5, -3, 2, -1, 1, 0, 1, 1, 2, 3, 5, 8, ...$, and in general $F_0 = 0$ and $F_{-n} = (-1)^{n+1} F_n$, where n is a positive integer. For more on the Fibonacci numbers, see chapter 2 and appendix B.

CHAPTER 2

$$\varphi$$

> Geometry has two great treasures: one
> is the theorem of Pythagoras; the other,
> the division of a line into extreme and
> mean ratio. The first we may compare to
> a measure of gold; the second we may
> name a precious jewel.
>
> **—JOHANNES KEPLER (1571–1630)**

CONSIDER A LINE SEGMENT AB OF LENGTH 1. We can divide
it internally by a point C in a variety of ways. For example, if C is the midpoint of AB, then $AC = CB = 1/2$ and so
$AB/AC = 2$. It seems the natural way of dividing AB; the
whole structure is perfectly symmetric and conveys a sense
of balance and stability.

The Greeks, however, thought of another way. Would it
be possible, they asked, to divide AB by a point C in such
a way that the whole segment is to the large part as the
large part is to the small? That is to say, $AB/AC = AC/CB$
(assuming that AC is the larger of the two parts). An equation of this type is called a *proportion*. The Greeks, to whom
ideas of harmony and symmetry were of great importance,
called this particular division an *extreme and mean proportion*; later it would be called the *golden section* (*sectio
aurea* in Latin), the *golden ratio*, or the *divine proportion*.
Of course, from a strictly mathematical point of view there
is nothing "divine" about this ratio, but the name stuck,
for reasons we'll soon see.

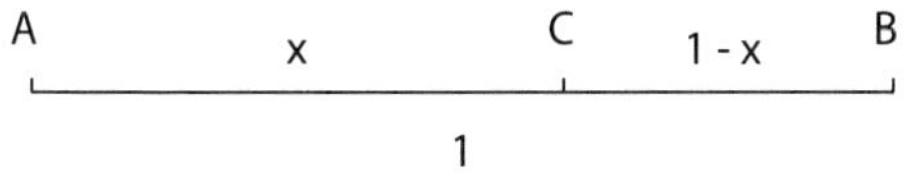

FIGURE 2.1. Dividing a line segment in the golden ratio

To translate the equation $AB/AC = AC/CB$ into something tangible, let the segment AB be of length 1 and denote the length of the larger part AC by x (figure 2.1). The smaller part is then $1 - x$, so our equation becomes

$$\frac{1}{x} = \frac{x}{1-x}. \tag{1}$$

Cross-multiplying the two sides gives us $x^2 = 1 - x$, or, after moving all terms to the left side,

$$x^2 + x - 1 = 0. \tag{2}$$

This is a quadratic equation whose coefficients are 1, 1, and −1. To solve it, we use the quadratic formula and get

$$x = \frac{-1 \pm \sqrt{1^2 - 4 \cdot 1 \cdot (-1)}}{2} = \frac{-1 \pm \sqrt{5}}{2}.$$

However, remembering that x denotes the length of a line segment and therefore cannot be negative, we choose only the + sign and get the single solution

$$x = \frac{-1 + \sqrt{5}}{2}. \tag{3}$$

With a calculator we can find the approximate value of this number, 0.618034. The exact decimal value, however, can never be found, because $\sqrt{5}$ is an irrational number whose decimal expansion neither terminates nor endlessly repeats (see frontispiece plate).

Now the number we are really looking for is not x but rather $1/x$, the numerical value of the ratio AB/AC. So while 0.618034 is still in the display of our calculator, we press the $1/x$ key and get the value 1.618034, which again is an approximation.

The keen reader has no doubt noticed that the values of x and $1/x$ as shown in the display have exactly the same digits after the decimal point; in fact, the two numbers differ exactly by 1. Is this just a curious coincidence? To find out, let us take the reciprocal of each side of equation (1):

$$x = \frac{1-x}{x}. \qquad (4)$$

Equations (1) and (4), of course, are entirely equivalent, and yet equation (4) has a slight advantage: because the minus sign appears in the numerator rather than the denominator, we can divide each term of the numerator by x and get

$$x = \frac{1}{x} - 1, \qquad (5)$$

which shows that the "curious coincidence" we discovered above is not a coincidence at all: it comes right out of the definition of the golden ratio.

In the following chapters, we'll need to use the *exact* value of $1/x$ (as opposed to its decimal approximation). We note that

$$\frac{\sqrt{5}-1}{2} \cdot \frac{\sqrt{5}+1}{2} = \frac{(\sqrt{5})^2 - 1^2}{4} = \frac{5-1}{4} = 1,$$

and therefore the expressions $\dfrac{\sqrt{5}-1}{2}$ and $\dfrac{\sqrt{5}+1}{2}$ are reciprocals of each other (notice again that they differ exactly by 1). Thus,

$$\frac{1}{x} = \frac{\sqrt{5}+1}{2}. \qquad (6)$$

This, then, is the celebrated number that became known as the golden ratio. Amazingly, it sneaks into many situations that seem to have little in common with one another: it shows up in nature, in art and architecture, and arguably even in music, and has been discussed endlessly by philosophers and art historians. But before we go into some of these appearances, we need to say a word about notation. You would think that such an important number should be given a universal symbol, just like π or e, but this is not so. Most mathematicians denote it by the Greek letter φ (phi, sometimes written ϕ), but some prefer the capitalized version Φ, while still others use the Greek τ (tau).[1] To add to the confusion, there is some disagreement whether any of these letters—and even the name *golden ratio*—should be given to $(1 + \sqrt{5})/2$ or to its reciprocal, $(-1 + \sqrt{5})/2$. We will go here with the majority and use the lower case φ to denote the number $\dfrac{1 + \sqrt{5}}{2}$. In view of this, we can rewrite equation (5) as $1/\varphi = \varphi - 1$, or equivalently as

$$\varphi = \frac{1}{\varphi} + 1. \qquad (7)$$

Equation (7) gives rise to a host of interesting results. For example, multiplying both sides by φ, we get $\varphi^2 = 1 + \varphi$. Multiplying again by φ, we get $\varphi^3 = \varphi + \varphi^2 = \varphi + (1 + \varphi) = 1 + 2\varphi$. Doing this repeatedly while each time replacing φ^2 by $1 + \varphi$, we get

$$\varphi^4 = 2 + 3\varphi, \quad \varphi^5 = 3 + 5\varphi, \quad \varphi^6 = 5 + 8\varphi, \quad \varphi^7 = 8 + 13\varphi, \ldots.$$

By now you may have recognized a pattern: each power of φ can be written as the sum of a number and a term involving just φ. And the coefficients of these expressions are none other than the Fibonacci numbers we met in chapter 1. In fact, if we denote the nth member of the Fibonacci sequence by F_n, we have the formula

$$\varphi^n = F_{n-1} + F_n \varphi. \tag{8}$$

And while we are dealing with powers of φ, here is one more interesting relation:

$$\varphi^n + \varphi^{n+1} = \varphi^{n+2}, \tag{9}$$

which mimics the definition of the Fibonacci numbers and holds true for any integer n, positive or negative.[2] These relations are just a few of the many connections between the Fibonacci numbers and the golden ratio (see appendix B for proofs of some of these relations).

The Greeks, who by all evidence were the first to discover the golden ratio, knew nothing of the Fibonacci numbers, nor were they versed enough in algebra to discover any of the relations mentioned above. They were, however, the undisputed masters of geometry, and they interpreted all numerical operations geometrically. For example, the sum $a + b$ of two numbers was interpreted as the total length of two line segments of length a and b placed end to end along the same line, while the product ab was the area of a rectangle with sides a and b. Each of these operations can be performed by using just two tools, a straightedge (an unmarked ruler) and a compass. According to tradition, it was the philosopher Plato (ca. 427–347 BCE) who first decreed that all geometric constructions should be done with just these two tools—the so-called Euclidean tools. Accordingly, we'll now show how to construct the golden ratio with the Euclidean tools.

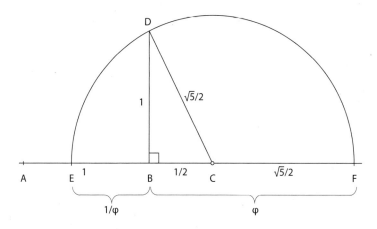

FIGURE 2.2. Constructing the golden ratio

Along a straight line, allocate a line segment *AB* of unit length and extend it indefinitely to the right (figure 2.2). On this extended line, allocate a segment *BC* of length 1/2, as well as a perpendicular *BD* of unit length. Connect points *C* and *D* by a straight line.

Using the Pythagorean theorem, we have

$$CD^2 = BC^2 + BD^2 = (1/2)^2 + 1^2 = 5/4,$$

So $CD = \sqrt{5}/2$. Now place the point of your compass at *C* and swing an arc of radius *CD*, meeting *BC* extended at *E* and *F*. We have

$$CE = CF = \sqrt{5}/2$$

and therefore

$$BF = BC + CF = 1/2 + \sqrt{5}/2 = \varphi,$$

and

$$EB = EC - BC = \sqrt{5}/2 - 1/2 = 1/\varphi.$$

Thus, a single construction gives us both the golden ratio and its reciprocal. We'll make use of this construction in chapter 4 to construct a regular pentagon.[3]

NOTES AND SOURCES

1. The notation φ for the golden ratio was first proposed in 1910 by James Mark McGinnis Barr (1871–1950), an American-British engineer, inventor, and polymath. He is said to have chosen the letter phi in honor of the Greek sculptor and architect Phidias (or Pheidias, ca. 480–430 BCE), who is said to have used the golden ratio in his work, although the claim is unsubstantiated.

    The first known decimal approximation of $1/\varphi$ was stated in 1597 by the German mathematician and astronomer Michael Maestlin (1550–1631) in a letter to Kepler; his value was about 0.6180340.

2. Equation (8) can be extended to negative powers of φ as well:

$$\varphi^{-1} = -1 + \varphi, \quad \varphi^{-2} = 2 - \varphi, \quad \varphi^{-3} = -3 + 2\varphi, \quad \varphi^{-4} = 5 - 3\varphi, \ldots,$$

    and in general $\varphi^{-n} = (-1)^n (F_{n+1} - F_n \varphi)$, where n is a positive integer.

    This formula can be simplified by rewriting the definition of the Fibonacci numbers as $F_n = F_{n+2} - F_{n+1}$ and extending it to nonpositive indices:

$$\ldots, -8, 5, -3, 2, -1, 1, 0, 1, 1, 2, 3, 5, 8, \ldots,$$

    and in general $F_0 = 0, F_{-n} = (-1)^{n+1} F_n$. Equation (8) then becomes $\varphi^{-n} = F_{-n-1} + F_{-n}\varphi$, which holds true for all integer values of n.

3. We skipped over the details of the elementary Euclidean constructions, such as transferring a line segment from one line to another, or erecting a perpendicular to a line from a given point on it. These constructions are given in appendix A.

Four Famous Irrational Numbers: $\sqrt{2}$, φ, e, and π

IT NEVER CEASES TO AMAZE ME that six of the most frequently used numbers in mathematics—0, 1, $\sqrt{2}$, φ, e, and π (in increasing numerical order)—fall within the first four units of the number line. Of these, the first two are at once integers and rational numbers, 0/1 and 1/1, but the others are irrational—they cannot be written as ratios of two integers. Consequently, their decimal expansion neither terminates nor endlessly repeats. In a sense, they are "infinite numbers," but they fall into two different classes of infinity: $\sqrt{2}$ and φ are *algebraic numbers*, whereas e and π are *transcendental numbers*. An algebraic number is a solution of a polynomial equation with rational coefficients. Thus, $\sqrt{2}$ and φ are the positive solutions of the polynomial equations $x^2 - 2 = 0$ and $x^2 - x - 1 = 0$, respectively. Transcendental numbers like e and π are not solutions of any polynomial equation with rational coefficients.[1]

The number line comprises infinitely many rational and irrational numbers, but not all infinities are the same. In a landmark work, Georg Cantor (1845–1918) in 1874 proved that whereas the set of all rational numbers is countable—its members can be put in a one-to-one correspondence with the set of the counting numbers—the set of all irrational numbers is *uncountable*—they cannot be put into such a correspondence. This means that the irrational numbers are actually more numerous than the rationals. Furthermore,

Cantor showed that the set of algebraic numbers is countable, but the set of transcendentals is uncountable. Not all infinities are alike.[2]

The rationals and irrationals also differ in another way: the former are closed under the four basic arithmetic operations, whereas the latter are not. That is, adding, subtracting, multiplying, or dividing (except by zero) two rational numbers always produces another rational number. The same is also true of any *finite* application of the four basic operations.[3] But not so with irrational numbers: the sum or product of two irrationals can result in a rational or an irrational number, as the following examples show: $\sqrt{2} \times \sqrt{2} = \sqrt{4} = 2$, a rational number, but $\sqrt{2} \times \sqrt{3} = \sqrt{6}$, an irrational number. In some cases, it is not even known what kind of number we get when adding two irrationals: for example, as of this writing it is not known whether $e + \pi$ is algebraic or transcendental.

Of the four famous irrational numbers, three have their origin in antiquity. A Babylonian clay tablet dating to roughly 1800–1600 BCE records, in cuneiform script and in sexagesimal (base 60) notation, the value of $\sqrt{2}$ as 1.414213, which is correct to the nearest one hundred thousandth.[4] And an Egyptian document, known as the Rhind Papyrus and dated to about the same period, shows how the Egyptians attempted to "square the circle"—to construct a square whose area is equal to that of a given circle. They devised a square of side 8/9 of the circle's diameter, leading to a value of π equal to 3.16049, in error of just 0.6 percent of the true value.[5] The golden ratio φ dates back to the sixth century BCE when the Pythagoreans discovered it and used it to construct the regular pentagon.

The number e, the base of natural logarithms, is of more recent vintage. It first appeared in the early seventeenth century after the invention of logarithms by John Napier. It is the limit of the expression $(1 + 1/n)^n$ as $n \to \infty$ and is equal to 2.71828, correct to five places.[6]

All four irrationals have strong connections to geometry. First, $\sqrt{2}$ is the length of the diagonal of a square of unit side; it is also the largest dimension of the square, its "diameter" of sorts. Similarly, φ is the length of the diagonal of a regular pentagon of unit side. The most famous of the four, the number π, is the circumference of a circle of unit diameter. Finally, e is at the heart of all exponential growth phenomena, represented graphically by the logarithmic spiral $r = e^{a\theta}$. All four numbers also appear in many other situations seemingly unrelated to geometry—among them algebra, probability, and analysis (the study of the continuum). And all four have remarkable infinite series or infinite product representations, some of which we mention in chapter 3.

In everyday use, the words *rational* and *irrational* mean "logical" and "illogical." That these words have also become part of the mathematical jargon is no accident. To the Pythagoreans, numbers were behind all of nature, from tiny atoms to the orbits of the planets: *Number rules the universe* was their motto, and by "number" they meant positive integers and their ratios. But one day a member of the Pythagorean sect by the name of Hippasus discovered that the square root of two cannot be expressed as a ratio of two integers. The discovery greatly bewildered them, and, according to legend, they threw him overboard the boat they were sailing in the Aegean Sea. Since the diagonal of a square of unit side has length $\sqrt{2}$, the Pythagoreans in effect regarded this length as a numberless quantity. To them it was illogical, irrational—and the name stuck.

★ ● ★

Four of the numbers mentioned above, plus the imaginary unit i (the square root of minus 1), are combined in one of the most famous equations in all of mathematics, $e^{\pi i} + 1 = 0$.

FIGURE S1.1. The eipiphiny by Philip Poissant

Discovered by the Swiss mathematician Leonhard Euler (1707–1783), it has become the subject of numerous interpretations, ranging from the divine to the banal. It plays a key role in trigonometry and in the theory of functions of complex variables (numbers of the form $x + iy$, where x and y represent real numbers).

Not surprisingly, Euler's equation has also caught the imagination of artists. Among them is my colleague Philip Poissant (b. 1947), a Canadian industrial designer who combined the five numbers in Euler's equation with the golden ratio φ in a miniature sculpture he called the *eipiphiny* (figure S1.1). It is a fitting tribute to one of the greatest mathematicians of all time.

NOTES AND SOURCES

1. Despite its spiritual connotation, the word *transcendental* in our context has nothing mysterious about it; it simply implies that numbers of this kind transcend (go beyond) the realm of polynomial equations with rational coefficients.
2. For a detailed survey of the various classes of infinities, see Eli Maor, *To Infinity and Beyond: A Cultural History of the Infinite* (Princeton, NJ: Princeton University Press, 2017), chs. 9–10.
3. This is not necessarily true when adding an *infinite* number of rationals, as the Gregory-Leibniz series $1 - 1/3 + 1/5 - 1/7 + - \cdots = \pi/4$ shows.
4. See Maor, *The Pythagorean Theorem: A 4,000-Year History*, ch. 1.
5. See Maor, *Trigonometric Delights*, p. 6.
6. See Maor, *e: The Story of a Number* (Princeton, NJ: Princeton University Press, 2009), ch. 1 and elsewhere.

CHAPTER 3

But Is It Divine?

The Golden Rectangle is said to be one of the
most visually satisfying of all geometric forms;
for years experts have been finding examples
in everything from the edifices of ancient
Greece to art masterpieces.

—DAVID BERGAMINI, *MATHEMATICS*,
LIFE SCIENCE LIBRARY (1965)

TAKE A LOOK AT THE RECTANGLE shown in figure 3.1. Aren't
you smitten by its elegant proportions, its divine beauty?
Well, maybe. You are looking at the "golden rectangle,"
whose length-to-width ratio is φ, the golden ratio. For generations, this rectangle has been praised by art critics,

FIGURE 3.1. The golden rectangle

architects, and aesthetic theorists as the one rectangle with the most pleasing proportions. But why? What makes this particular rectangle seem so pleasing to the eye? Is it really as beautiful as claimed by the experts?

Perhaps the most famous painting in the history of art is Leonardo da Vinci's *Mona Lisa*, attracting huge crowds to the Louvre in Paris to get a glimpse of ... themselves, as the painting has been covered by protective glass that reflects the image of the visitor more than it shows the legendary smiling lady. Numerous attempts have been made to discover the golden ratio in this painting, as shown in figure 3.2. The picture shows several golden rectangles superimposed on the famous lady's image, all with length-to-width ratio of 1.618. But look carefully: How are these rectangles defined, really? For instance, the largest rectangle goes down from her chin to her right-hand thumb, but what about the rectangle's width? It is not clear at all how it is defined, and, to be honest, you get the feeling that it has been put there only to produce the purported ratio. And what is the purpose of showing it in the first place? To convince us that da Vinci had this particular ratio in mind when he drew the painting? There is not a shred of evidence in Leonardo's writings (many of which have survived) to prove that this was his intention.

Well, you might say, surely some ratios in that painting, if not exactly equal to the golden ratio, are close enough to it. But once you start dealing with approximations, the rules become rather vague. For example, three is a fairly close approximation to π; in fact, it differs from π by only 4.5 percent. But if we take this close approximation too seriously, we'll find a connection to π whenever three shows up—that is to say, *everywhere*.

Or take the Parthenon temple on top of the Acropolis hill in Athens. Numerous diagrams have been created purportedly showing that this most famous relic of ancient

FIGURE 3.2. Leonardo da Vinci, *Mona Lisa*

FIGURE 3.3. The Parthenon

Greece has the proportions of the golden ratio (figure 3.3). It has been claimed that the Greeks designed the temple specifically in this proportion. But look carefully: the horizontal lines can be drawn at several locations on the temple's facade, giving you slightly different proportions. Yes, they are all close to φ—close, but not equal. Some obsessed believers would even adorn the temple with rectangles that do indeed have the golden proportion but do not point to any specific feature of the building's facade (see the leftmost rectangle in figure 3.3). And in any case, no blueprint of the building has survived—if there ever was one—so all the aesthetic theories behind this revered relic are based on sheer speculation, unsupported by hard evidence. Yet golden ratio aficionados, like circle squarers, remain undeterred, always finding fresh "evidence" to support their claim. And so the myth of the divine section keeps perpetuating itself, feeding on its own momentum.

FIGURE 3.4. Salvador Dalí, *The Sacrament of the Last Supper*

Closer to our time, several artists have indeed created works specifically based on the golden ratio. Perhaps the most famous of these is *The Sacrament of the Last Supper* by Spanish artist Salvador Dali (1904–1989). Dali created it in 1955 during his "post–World War II period," when he became obsessed with ideas taken from science and religion. In 1948 Dali met the Romanian-born polymath Matila Costiescu Ghyka (1881–1965), an eccentric naval officer of royal descent who became a mathematician, historian, philosopher, diplomat, and novelist. In 1946 Ghyka published a small book, *The Geometry of Art and Life*, containing numerous photographs and hand-drawn diagrams purporting to show the role of the golden ratio in art and architecture, in the natural world, and in anatomy. The slim volume influenced a whole cadre of artists and naturalists. Among them was Dali, who created his huge oil painting with several allusions to the golden ratio (figure 3.4). The scene of the Last Supper takes place inside what appears to be a dodecahedron, whose twelve pentagonal faces may allude to the twelve apostles (although only four faces are partially shown), but they also remind us that the very construction of the pentagon rests on the golden ratio. What's more, Dali drew this enormous work in a rectangle of dimensions 166.7×267 cm ($65\frac{5}{8} \times 105\frac{1}{8}$ in.), whose aspect ratio is 1.6017, just about 1 percent short of 1.6180.

Perhaps the most bizarre association of the golden ratio with a work of art was made in connection with music. In a long article, "Duality and Synthesis in the Music of Béla Bartok," the author, Ernö Lendvai, goes into a measure-by-measure analysis of Bartok's *Sonata for Two Pianos and Percussion* to uncover the hidden existence of Fibonacci numbers and the golden ratio in the sequence of notes and their rhythmic structure.[1] Perhaps there is such a connection, or perhaps there isn't. In any case, we have to bear in mind that aural perception—which music is all about—is entirely different from spatial, visual perception, from which the golden ratio originates.

For golden ratio fans, I mention two more cases: Have you ever measured the dimensions of your credit card? In case you haven't, here they are: 85.60×53.98 mm, giving an aspect ratio of 1.586, rounded to three decimals. That's just 1.57 percent off the golden ratio! Is this by design? By all likelihood, the answer is no.

And if you are, like myself, a metric person, you've no doubt converted miles to kilometers more times than you care to remember: 1 statute mile = 1.60934 km. Aha—the golden ratio again! Well, almost....[2]

The first mention of what is now called the golden ratio is in Euclid's *Elements*, written around 300 BCE in Alexandria. Arranged in thirteen parts ("books"), it is a compilation of geometry and number theory as it was known in his time, organized in a logical order beginning with the most elementary concepts, such as a point and a line. It contains ten axioms (called "postulates" and "common notions"), numerous definitions, and 465 theorems ("propositions"), complete with their proofs ("demonstrations"). Its strict definition-axiom-theorem-proof organization has become the model for generations of mathematicians of how mathematics should be practiced.

Proposition 30 of book VI says: "To cut a given finite straight line in extreme and mean ratio."[3] Euclid shows us a way to construct this mean, and he uses it later in Book XIII to construct the face of a regular dodecahedron. His treatment is entirely geometric, with no reference to the numerical value of this mean. Certainly, there's no hint of any "divine" qualities to this ratio. That had to wait a little while—until 1509, to be exact—when a Franciscan friar and mathematician by the name of Luca Pacioli (ca. 1445–1517[4]) published a geometry book, *De divina proportione*, which referred to the golden ratio as a symbol of divinity, of God. It is likely, however, that his divine attribution had more to do with the fact that φ is an irrational number than with its supposed aesthetic appeal. Pacioli struck a friendship with the young Leonardo da Vinci, and da Vinci took mathematics lessons from his senior friend, for whom he illustrated the book with drawings of the Platonic solids (see page ii).

The obsession with the artistic significance of the golden ratio was actually a product of the aesthetic movement of the nineteenth century. As far as is known, the adjective *golden* was first used by German mathematician Martin Ohm (1792–1872), brother of the more famous physicist Georg Ohm. In his book *Die reine Elementar-Mathematik* (*Pure Elementary Mathematics*, 1835) he called φ the "golden section" (*goldener Schnitt* in German), and the name stuck.[5]

★　⬠　★

To my mathematical eyes, there are indeed three features that justify calling this ratio "divine," but they have little to do with artistic aesthetics. The first involves two formulas that leave us in awe of the beauty of pure mathematics. Start with equation (7) in chapter 2, $\varphi = 1/\varphi + 1$, and multiply it by φ:

$$\varphi^2 = 1 + \varphi. \tag{1}$$

Now take the positive square root of both sides:

$$\varphi = \sqrt{1 + \varphi}. \tag{2}$$

This is a somewhat unusual way to deal with an equation, as we normally want to have the unknown all by itself on one side of the equation and a known quantity on the other side. But now that we've written down equation (2), let's do something even more unusual: let's plug the φ on the left side back into the right side, giving us $\varphi = \sqrt{1 + \sqrt{1 + \varphi}}$. Doing this again and again—in essence, recycling φ into its own equation—we get

$$\varphi = \sqrt{1 + \sqrt{1 + \sqrt{1 + \cdots + \sqrt{1 + \varphi}}}}.$$

Repeating this process indefinitely, the φ inside the innermost radical sign keeps being pushed further and further to the right, giving us the remarkable limiting expression

$$\varphi = \sqrt{1 + \sqrt{1 + \sqrt{1 + \cdots}}}. \tag{3}$$

But wait: How do we know that this infinite chain of nested square roots will converge—and indeed to the value of φ? We cannot assume that the φ inside the innermost radical will just go away. This is one of those tricky formulas that intrigued the seventeenth- and eighteenths-century mathematicians, who tried to deal with infinity as if it were an ordinary number. So to justify this formula, we offer here a geometric demonstration (see figure 3.5). On a rectangular coordinate system, plot the graphs of $y = x$ (a straight line through the origin at 45 degrees to either axis) and $y = \sqrt{1 + x}$ (one half of a horizontal parabola shifted one unit to the left). Start at $x = 1$ and move up until you hit the parabola at $(1, \sqrt{2})$. Now move horizontally to the line $y = x$ and then up again to the parabola, reaching it at the point $(\sqrt{2}, \sqrt{1 + \sqrt{2}})$. Repeat this process; the points on the two graphs follow a staircase path and merge where the graphs intersect—that is, where $x = \sqrt{1 + x}$, or $x^2 = 1 + x$. Rewriting this equation as $x^2 - x - 1 = 0$ and using the quadratic formula, we get $x = \dfrac{1 + \sqrt{5}}{2}$, the golden ratio (we discard the negative solution $\dfrac{1 - \sqrt{5}}{2}$, as it lies outside the range of the parabola). The process also shows that the convergence to this limiting value is extremely fast.

It is indeed surprising that the simple-looking expression $\sqrt{1 + \sqrt{1 + \sqrt{1 + \ldots}}}$, involving only the number 1, should converge to the value $(1 + \sqrt{5})/2 \sim 1.618$. But if you are not persuaded, try to calculate the first few terms on your calculator; let's call them φ_n, where n denotes the number of nested radicals. We have the recurrence formula $\varphi_0 = 0$, $\varphi_{n+1} = \sqrt{1 + \varphi_n}$, $n = 0, 1, 2, 3, \ldots$, giving us the values

$$\varphi_1 = \sqrt{1} = 1, \quad \varphi_2 = \sqrt{1 + \sqrt{1}} = \sqrt{2} = 1.414,$$
$$\varphi_3 = \sqrt{1 + \sqrt{2}} = 1.554, \ldots,$$

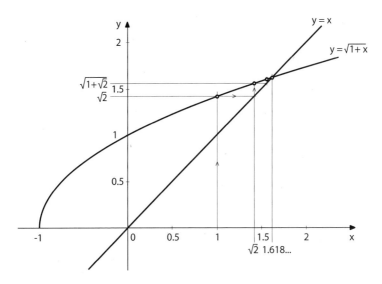

FIGURE 3.5. Geometric demonstration of $\varphi = \sqrt{1 + \sqrt{1 + \sqrt{1 + \cdots}}}$

all rounded to three decimal places. After six more iterations, we get the correct value, again rounded to three places. A programmable calculator, of course, will do these iterations by itself.

★ ⬠ ★

For the second "divine" feature, we start again with the equation $\varphi = 1 + 1/\varphi$ and plug in the φ on the left side into the right side, getting $\varphi = 1 + \dfrac{1}{1 + 1/\varphi}$. Doing this repeatedly, we get the continued fraction

$$\varphi = 1 + \cfrac{1}{1 + \cfrac{1}{1 + \cfrac{1}{1 + \cdots}}}, \tag{4}$$

a formula just as remarkable in its simplicity as equation (3). But behind the endlessly recurring string of 1s lurks a surprise. To discover it, we again write the formula as a recurrence formula, $\varphi_{n+1} = 1 + 1/\varphi_n$, where n stands for the number of leading 1s. We have

$$\varphi_1 = 1, \quad \varphi_2 = 1 + 1/1 = 2, \quad \varphi_3 = 1 + 1/2 = 3/2 = 1.5,$$
$$\varphi_4 = 1 + 1/(3/2) = 1 + 2/3 = 5/3 = 1.667,$$
$$\varphi_5 = 1 + 1/(5/3) = 1 + 3/5 = 8/5 = 1.6,$$
$$\varphi_6 = 1 + 1/(8/5) = 1 + 5/8 = 13/8 = 1.625,$$

and so on. Do these numbers look familiar? Why, they are none other than the ratios of successive Fibonacci numbers to their predecessors. The next four iterations produce the fractions

$$21/13 = 1.615, \quad 34/21 = 1.619, \quad 55/34 = 1.618,$$
$$89/55 = 1.618,$$

all rounded to three places. We see that after the ninth iteration the values settle on 1.618, the golden ratio. Note that the fractions approach this limiting value alternatively from above and below—unlike the sequence of nested square roots, which approach the limit steadily from below. Note also that all the approximations generated by equation (4) are rational numbers (that is, fractions), again in contrast to equation (3), whose approximations—except for the first one—are all irrational numbers. This is perhaps one of the most remarkable features of mathematics—that an infinite sequence of rational numbers may converge to an irrational number.[6]

Beyond the sheer simplicity of equations (3) and (4), they establish a connection between two concepts taken from entirely different branches of mathematics: the Fibonacci numbers, which squarely belong to the realm of arithmetic, and the golden ratio, whose origin is in geometry. It is this connection, I think, that provides the ultimate justification for the title *divine proportion*.

It is not known who discovered equations (3) and (4). But we do know the name of the next player in this story: Simon Jacob (ca. 1510–1564), a German mathematician who in 1560 wrote a book on the art of reckoning. Not much is known about him, but he seems to have been the first to notice that the ratio of a Fibonacci number to its predecessor approaches the golden ratio as we steadily move up the sequence (for a proof, see appendix B). This discovery is usually credited to Johannes Kepler (1571–1630), considered the father of modern astronomy but also a mathematician and naturalist (his many writings include a book on snowflakes). Kepler was an ardent Pythagorean who believed in the mystic power of numbers and shapes, and he was fascinated by the division of a line into extreme and mean proportion (see the epigraph to chapter 2). So Kepler gets the credit for popularizing the golden ratio and making it the subject of ongoing research.

I kept the third reason for endowing φ with the title *divine* to the end: it turns out to be the key to constructing the regular pentagon with Euclidean tools. It is this construction that is the subject of our next chapter.

NOTES AND SOURCES

1. Gyorgy Kepes, ed., *Module, Symmetry, Proportion*, pp. 174–93.
2. For more on the fads and myths surrounding the golden ratio, see Mario Livio, *The Golden Ratio*, ch. 3.
3. Euclid, *The Elements*, vol. 2, pp. 267–68.
4. There is some disagreement as to Pacioli's dates of birth and death; some authors give them as 1447–1517 or 1445–1514.
5. Lynn Gamwell, *Mathematics +Art*, pp. 91–107.
6. Another famous example of this is the Mercator series, $1 - 1/2 + 1/3 - 1/4 + - \cdots$ $= ln2$, where $ln2$ is the natural logarithm of 2.

CHAPTER 4

Constructing the Regular Pentagon

Euclid's work will live long after all the text-books of the present day are superseded and forgotten. It is one of the noblest monuments of antiquity.

—H.S.M. COXETER, *INTRODUCTION TO GEOMETRY* (1961)

PROPOSITION 10 OF BOOK IV of Euclid's *The Elements* says, "To construct an isosceles triangle having each of the angles at the base double of the remaining one."[1] Hidden behind this innocuous statement lies the secret of constructing the regular pentagon. To see this, let us rephrase the proposition in modern language: To construct an isosceles triangle with top angle α and each of the base angles 2α. This means $5\alpha = 180°$, so $\alpha = \dfrac{180°}{5} = 36°$, that is, one-tenth of a full rotation. Thus, ten of these triangles, joined side by side at their common vertex, would form a regular *decagon*—a ten-sided polygon with all its sides and all its angles equal. From this it is just one more step to get a regular pentagon—all we need to do is join every other vertex with a straight line, and our construction is complete (figure 4.1).

Well, not so fast. First of all, the early Greeks did not use our degree-minute-second system of angular measure; their only angular unit was a right angle, and they called it as such, not 90 degrees. Thus, $36° = (2/5) \times 90°$, or two-fifths

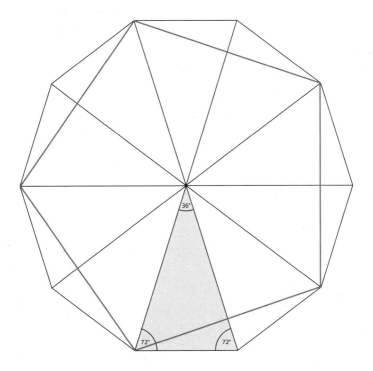

FIGURE 4.1. Constructing a regular pentagon I

of a right angle. And second, they did not use a single letter like α to denote the measure of an angle, but rather the cumbersome notation ACB for the angle between two rays CA and CB joined at point C.

Euclid's challenge in proposition IV 10 is to construct this triangle with Euclidean tools. He gives the procedure, which relies directly on the ability to find the point that divides a line segment in an extreme and mean proportion. This is followed by a proof ("demonstration") confirming that the construction does indeed produce the required triangle. The proof is rather long and can easily wear down a modern reader, with its multiple triple-letter designation of angles (figure 4.2); one has to patiently plod through it, making sure not to confuse, say, angle

PROPOSITION 10.

To construct an isosceles triangle having each of the angles at the base double of the remaining one.

Let any straight line AB be set out, and let it be cut at the point C so that the rectangle contained by AB, BC is equal to the square on CA;   [II. 11]

with centre A and distance AB let the circle BDE be described,

and let there be fitted in the circle BDE the straight line BD equal to the straight line AC which is not greater than the diameter of the circle BDE.   [IV. 1]

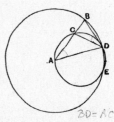

Let AD, DC be joined, and let the circle ACD be circumscribed about the triangle ACD.
[IV. 5]

Then, since the rectangle AB, BC is equal to the square on AC,

and AC is equal to BD,

therefore the rectangle AB, BC is equal to the square on BD.

And, since a point B has been taken outside the circle ACD,

and from B the two straight lines BA, BD have fallen on the circle ACD, and one of them cuts it, while the other falls on it,

and the rectangle AB, BC is equal to the square on BD,

therefore BD touches the circle ACD.   [III. 37]

Since, then, BD touches it, and DC is drawn across from the point of contact at D,

therefore the angle BDC is equal to the angle DAC in the alternate segment of the circle.   [III. 32]

Since, then, the angle BDC is equal to the angle DAC, let the angle CDA be added to each;

therefore the whole angle BDA is equal to the two angles CDA, DAC.

FIGURE 4.2. Constructing the golden triangle (Euclid IV 10)

But the exterior angle BCD is equal to the angles CDA, DAC;   [I. 32]

therefore the angle BDA is also equal to the angle BCD.

But the angle BDA is equal to the angle CBD, since the side AD is also equal to AB;   [I. 5]

so that the angle DBA is also equal to the angle BCD.

Therefore the three angles BDA, DBA, BCD are equal to one another.

And, since the angle DBC is equal to the angle BCD,

the side BD is also equal to the side DC.   [I. 6]

But BD is by hypothesis equal to CA;

therefore CA is also equal to CD,

so that the angle CDA is also equal to the angle DAC;   [I. 5]

therefore the angles CDA, DAC are double of the angle DAC.

But the angle BCD is equal to the angles CDA, DAC;

therefore the angle BCD is also double of the angle CAD.

But the angle BCD is equal to each of the angles BDA, DBA;

therefore each of the angles BDA, DBA is also double of the angle DAB.

Therefore the isosceles triangle ABD has been constructed having each of the angles at the base DB double of the remaining one.

Q. E. F.

There is every reason to conclude that the connexion of the triangle constructed in this proposition with the regular pentagon, and the construction of the triangle itself, were the discovery of the Pythagoreans. In the first place the Scholium IV. No. 2 (Heiberg, Vol. v. p. 273) says "this Book is the discovery of the Pythagoreans." Secondly, the summary in Proclus (p. 65, 20) says that Pythagoras discovered "the construction of the cosmic figures," by which must be understood the five regular solids. Thirdly, Iamblichus (*Vit. Pyth.* c. 18, s. 88) quotes a story of Hippasus, "that he was one of the Pythagoreans but, owing to his being the first to publish and write down (the construction of) the sphere arising from the twelve pentagons (τὴν ἐκ τῶν δώδεκα πενταγώνων), perished by shipwreck for his impiety, having got credit for the discovery all the same, whereas everything belonged to HIM (ἐκείνου τοῦ ἀνδρός), for it is thus that they refer to Pythagoras, and they do not call him by his name." Cantor has (I₃, pp. 176 sqq.) collected notices which help us to form an idea how the discovery of the Euclidean construction for a regular pentagon may have been arrived at by the Pythagoreans.

Plato puts into the mouth of Timaeus a description of the formation from

FIGURE 4.2. *Continued.*

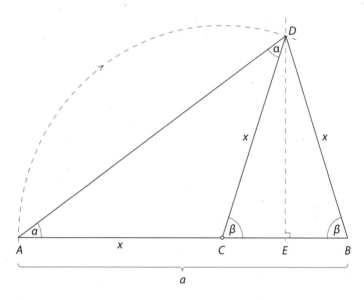

FIGURE 4.3. Constructing the golden triangle (modern notation)

ABC with angle *BAC*. We give here the construction and proof, both rephrased in modern notation and using the term *golden ratio* instead of the cumbersome "extreme and mean proportion."

Figure 4.3 shows a line segment *AB* of length *a* divided in the golden ratio by point *C*, using the construction described in chapter 2. Let $AC = x$. Let *E* be the midpoint of *CB*. Erect the perpendicular bisector to *CB* at *E*; it is the locus of all points equidistant from *C* and *B*. Now place the point of your compass at *C* and swing an arc of radius $CA = x$, cutting the perpendicular bisector at *D*. Connect each of the points *A*, *C*, and *B* with *D* to form two isosceles triangles, *ACD* and *CBD* (the latter because $EB = EC$) and the larger triangle *ABD*. We claim that *ABD* is the required triangle.

Proof: Denote the internal angles at *A* and *B* by α and β, respectively. Since triangle *ACD* is isosceles, its internal angle at D is also α. But triangle CBD is also isosceles, and therefore its internal angle at C is also β. We now find the length of AD, using the Pythagorean theorem twice:

$$
\begin{aligned}
AD^2 &= AE^2 + ED^2 = (AC + CE)^2 + (CD^2 - CE^2) \\
&= AC^2 + 2AC \cdot CE + CD^2 \\
&= x^2 + 2x \cdot (a - x)/2 + x^2 \\
&= x^2 + ax.
\end{aligned} \tag{1}
$$

But remember, point C divided segment AB in the golden ratio, so $AB/AC = AC/CB$, or $AB \cdot CB = AC^2$; that is, $a \cdot (a - x) = x^2$, or $x^2 + ax = a^2$. Therefore, replacing the right side of equation (1) by a^2, we get $AD^2 = a^2$ and finally $AD = a$. But this shows that triangle ABD too is isosceles, and therefore $\angle ADB = \beta$. Finally, $\angle BCD = \beta$ is external to triangle ACD and therefore is equal to the sum of its two nonadjacent angles; that is, $\beta = 2\alpha$. Thus, we have constructed an isosceles triangle ABD with each of its base angles equal to twice the vertex angle: QED.[2]

We can now determine the measure of the angles of triangle ABD: we have $\beta = 2\alpha$ and $\alpha + 2\beta = 180°$, from which we get $\alpha = 36°$ and $\beta = 72°$ (note also that line CD bisects angle ADB). This is all we need to construct a regular pentagon of arbitrary size, as described in this chapter's opening paragraph.

★　⬟　★

The 72°-72°-36° triangle became known as the *golden triangle*, and it plays a central role in every aspect of the regular pentagon. Since most modern textbooks start with this triangle as if it were already given and then proceed to show its properties, we now present this approach as a review of sorts; but keep in mind that it is the exact reverse of Euclid's approach. We now show that the side-to-base ratio of the golden triangle is exactly φ, the golden ratio.

Let the triangle be ABC, with the 36° angle at the top vertex C (figure 4.4a). Bisect the 72° angle at A and extend the bisector until it meets the opposite side BC at D. This produces two isosceles triangles, the 72°-72°-36° triangle BDA and the 36°-36°-108° triangle ACD. Of these, the former is similar to triangle ABC, so we have

$$\frac{AB}{BC} = \frac{BD}{DA}.$$

Setting $AB = 1$ and $BC = x$, we have $AB = AD = DC = 1$, $AC = BC = x$, and $BD = x - 1$. We thus get

$$\frac{1}{x} = \frac{x-1}{1},$$

which leads to the quadratic equation $x^2 - x - 1 = 0$. Solving it and taking only the positive solution (because x denotes length, which cannot be negative), we get

$$x = \frac{1 + \sqrt{5}}{2},$$

that is, the golden ratio φ. To recap: *In a golden triangle, the side-to-base ratio is φ.* In what follows, we'll frequently make use of this fact.

Before moving on, let us mention a remarkable property of the golden triangle: it can be split into two triangles, the smaller of which is similar to the original triangle and is therefore also a golden triangle. Since triangles ABC and BDA are similar, we can dissect triangle BDA in the same way as we did with triangle ABC to get a still smaller golden triangle embedded in it. In fact, we can repeat the process again and again, getting an infinite progression of ever-smaller golden triangles, each nested inside its predecessor. Fig. 4.4b, by Eugen Jost, shows this process graphically.

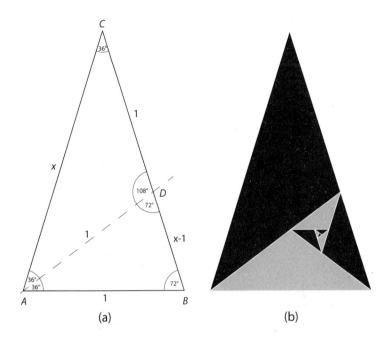

FIGURE 4.4. Dissecting a golden triangle

★ ⬠ ★

The construction we discussed above applies to a regular pentagon of arbitrary size. More often, however, we may be asked to construct a pentagon with a given dimension, such as the length of its side or the radius of its circumscribing circle. Let us begin with constructing a regular pentagon of side $AB = 1$ (figure 4.5). Extend AB to the right and allocate on it a segment BQ equal in length to φ, using the construction shown in figure 2.2. Place your compass at B and swing an arc of radius $BQ = \varphi$. Then do the same with your compass placed at A (the two arcs are shown as dashed in figure 4.5). The two arcs intersect at D. Next, from each of the points A and D, swing an arc of radius 1 (not shown in the figure); the two arcs intersect at E.[3] Now do the same

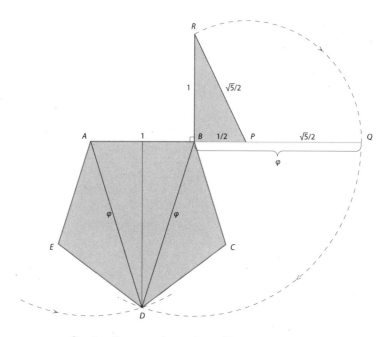

FIGURE 4.5. Constructing a regular pentagon II

from points B and D, producing point C. Connect points A, B, C, D, and E, and your pentagon is complete.[4]

This procedure relies on our earlier construction of the golden ratio, which in turn is based on the expression $(1 + \sqrt{5})/2$; but this is not how Euclid did it. In fact, *The Elements* does not give us an explicit recipe for constructing a regular pentagon. Proposition IV 10 only tells us how to construct a golden triangle, leaving it to the reader to figure out how to go from there to the pentagon (see again the opening paragraph of this chapter).

Proposition IV 10 is followed by three theorems that tell us how to construct a regular pentagon in relation to a circle. Thus, proposition IV 11 says, "In a given circle to

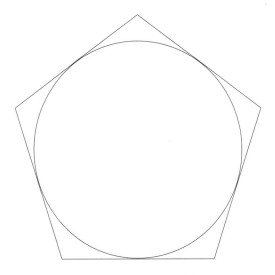

FIGURE 4.6. A pentagon circumscribed about a given circle, or a circle inscribed inside a given pentagon

inscribe an equilateral and equiangular pentagon." This is followed by propositions IV 12 ("About a given circle to circumscribe an equilateral and equiangular pentagon") and IV 13 ("In a given pentagon, which is equilateral and equiangular, to inscribe a circle"). In the last two, the circle is tangent to each side of the pentagon at its midpoint (figure 4.6); the difference is that in proposition IV 12 the *circle* is given, while in proposition IV 13 it is the pentagon that is given; in a sense, the two problems are inverses of each other.

Euclid's procedure in IV 11 is again strictly geometric: first, he instructs us to create a golden triangle of arbitrary size, then inscribe in the given circle a triangle with the same angles as the first triangle; that is, another golden triangle (figure 4.7). Let this second triangle be ACD, with the top vertex A (of 36° in modern notation) placed at any point on the circle. Then bisect the angles at C and D, and let the bisectors meet the circle at E and B, respectively. Join points A, B, C, D, and E by straight lines, and you get at once

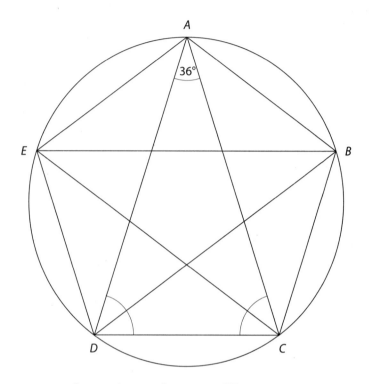

FIGURE 4.7. Constructing a regular pentagon III

a regular pentagon and its associated pentagram (a five-sided "star pentagon"), both inscribed in the given circle.[5]

Euclid's construction, however, is moot on one point: How do you fit a golden triangle of unspecified size in a given circle? It can be done, but it complicates things considerably.[6]

We give now two alternative methods for inscribing a pentagon in a given circle, following a more modern approach. But first we must find the relation between the side of an inscribed pentagon and the radius of the circumcircle. Figure 4.8 shows a golden triangle *ABC* inscribed in a circle

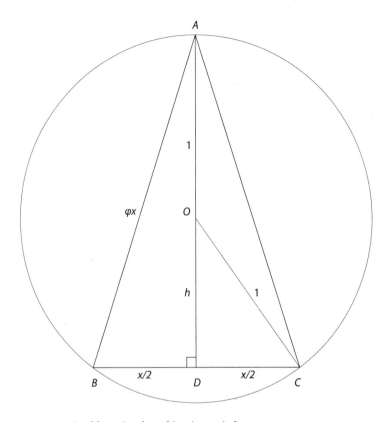

FIGURE 4.8. A golden triangle and its circumcircle

with center O and radius $OC = 1$. Drop the perpendicular OD from O to the base BC and denote its length by h. Let the length of BC be x. Remembering that $AC = \varphi x$ (because ABC is a golden triangle, whose side-to-base ratio is φ) and using the Pythagorean theorem twice, first for triangle CDO and again for triangle CDA, we have

$$(x/2)^2 + h^2 = 1^2 \quad \text{and} \quad (x/2)^2 + (1+h)^2 = (\varphi x)^2.$$

Solving this pair of equations for h and x is a bit tedious though not particularly difficult, and I leave it to the reader (it helps to convert powers of φ into linear expressions of

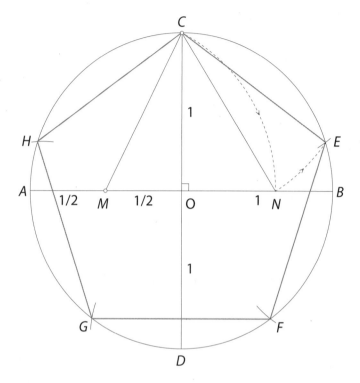

FIGURE 4.9. Inscribing a pentagon in a circle: alternative method I

the form $a + b\varphi$, as we saw in chapter 2). We give here the answer for x:

$$x = \sqrt{3 - \varphi} = \sqrt{\dfrac{5 - \sqrt{5}}{2}}.$$

We are now ready to tackle the two alternative construction methods.

Method I: Let the circle have center O and radius 1 (figure 4.9). Draw two perpendicular diameters AB and CD and bisect the radius OA at M. Connect M and C, place the point of your compass at M, and swing an arc of radius MC, cutting AB at N. Connect N and C. We now show that segment NC is equal to the side x of the required pentagon.

Proof:

$$MC^2 = MO^2 + OC^2 = (1/2)^2 + 1^2 = 5/4,$$

so

$$MC = \sqrt{5}/2 = MN.$$
$$NC^2 = NO^2 + OC^2 = (MN - MO)^2 + OC^2$$
$$= (\sqrt{5}/2 - 1/2)^2 + 1^2 = (\sqrt{5} - 1)^2/4 + 1.$$

But this last expression is equal to $(5 - \sqrt{5})/2$, as can easily be seen by doing the algebra. Thus,

$$NC = \sqrt{\frac{5 - \sqrt{5}}{2}},$$

which is precisely the length of one side of a regular pentagon inscribed in a unit circle.

But we still have to make NC a chord of the circle. To this end, place your compass at C and swing an arc of radius CN, cutting the circle at E. Repeat the process with your compass placed at E, cutting the circle at F. Do this two more times, producing points G and H. Connect C to E, E to F, and so on with straight lines, and your pentagon $CEFGH$ is complete. As a check on your accuracy, if you place your compass at H and swing an arc as before, it should cut the circle exactly at C.

Note: All the involved constructions, such as drawing two perpendicular diameters or finding the midpoint of the radius, involve elementary Euclidean constructions, and we omit them here. Also, in the most restrictive interpretation of the phrase "a given circle," the center and radius are *not* given but must be found by elementary constructions. For details, see appendix A.

Method II: Starting again with the unit circle and center at O, draw two perpendicular diameters AB and CD (figure 4.10). As before, bisect OA at M and connect M with C.

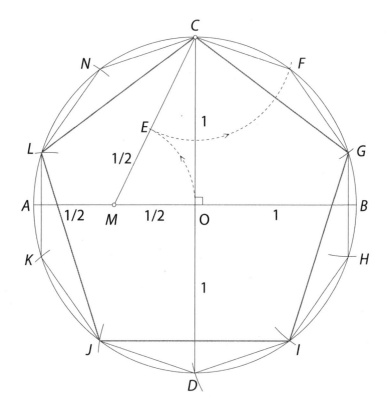

FIGURE 4.10. Inscribing a pentagon in a circle: alternative method II

Place the point of your compass at M and swing an arc of radius $MO = 1/2$, cutting MC at E. We now show that CE is equal in length to the side of an inscribed regular *decagon*.

Proof: As before, $MC = \sqrt{5}/2$, so $EC = MC - ME = \sqrt{5}/2 - 1/2 = (-1 + \sqrt{5})/2$. But this last expression is exactly $1/\varphi$ (see chapter 2), which is the base-to-side ratio of a golden triangle. In order to place this segment as a chord of the circle, put the point of your compass at C and swing an arc of radius CE, cutting the circle at F. Triangle CFO (only two sides of which are shown in the figure) is therefore a golden triangle with a vertex angle of 36° at O, so CF is the side of the decagon. Now repeat the process with your compass placed at F to produce the next vertex G, and so on until you arrive back at C. Finally, to get the regular *pentagon*, connect every other vertex, and you are done.[7]

This construction has an interesting history. Sir Thomas Heath, editor of the modern English edition of *The Elements*, claims that it was proposed by H. M. Taylor. Not much is known about this person, and a Google search found only three short articles about him. Henry Martyn Taylor (1842–1927) was an English barrister and mathematician, whose interest was in intuitive geometry. At the age of fifty-two, he was stricken by blindness; but far from hindering him in his career, he chose to make the best of it. Here is a quote from the Wikipedia article about him:

> He devised a Braille notation when he was overtaken by blindness in 1894, when engaged in the preparation of an edition of Euclid for the Cambridge University Press. By means of his ingenious and well thought out Braille notation he was enabled to transcribe many advanced scientific and mathematical works, and in 1917, with the assistance of Mr. Emblen, a blind member of the staff of the National Institute for the Blind, he perfected it. It was recognised as so comprehensive that it was soon adopted as the standard mathematical and chemical notation, and is universally used by English-speaking people.

The article goes on to say that Taylor was elected mayor of Cambridge for 1900–1901 and fellow of the Royal Society in 1898. He died in Cambridge and is buried at the Parish of Ascension Burial Ground there.[8]

This is not the first instance of a blind person making mathematical contributions. The great Leonhard Euler spent his last seventeen years completely blind, but this did not slow down his mathematical creativity, and he kept active to his last day. We also have the case of Emma A.

Coolidge (born in 1857 in Sturbridge, Massachusetts), a blind woman who in 1888 devised a dissection proof of the Pythagorean theorem. We can only be awed by these cases of mind's triumph over body.[9]

NOTES AND SOURCES

1. Euclid, *The Elements*, vol. 2, pp. 96–97.
2. QED stands for *quod erat demonstrandum*, "that which was to be demonstrated."
3. For the two arcs to meet, each radius must be greater than half the segment. Indeed, $AD = BD = BQ = (1 + \sqrt{5})/2 \sim 1.618$, so $AE = 1 > AD/2 \sim 0.809$, ensuring that the arcs meet.
4. The text and construction are adapted from the authors' book *Beautiful Geometry*, pp. 177–79.
5. Euclid, *The Elements*, vol. 2, pp. 100–101.
6. To start with, choose any point on the circle, say, point A, and draw the circle's diameter through A. Next, bisect the top angle of a golden triangle of arbitrary size, then copy each of the half angles (18°) on either side of the diameter at A. The two outer sides meet the circle at points B and C. Connect B and C, and you have the golden triangle inscribed in the given circle.
7. There are still other ways to construct a regular pentagon under various restricting conditions, including a Mascheroni construction using a compass only, and an approximate construction attributed to Albrecht Dürer. I refer the reader to the following articles (all taken from the excellent website Cut the Knot, https://www.cut-the-knot.org, whose emblem is a regular pentagon):
   - ♠ "Approximate Construction of Regular Pentagon" by A. Dürer
   - ♠ "Construction of Regular Pentagon" by H. W. Richmond
   - ♠ "Inscribing a Regular Pentagon in a Circle—and Proving It"
   - ♠ "Regular Pentagon Construction" by Y. Hirano
   - ♠ "Regular Pentagon Inscribed in Circle by Paper"
   - ♠ "Mascheroni Construction of a Regular Pentagon"
   - ♠ "Regular Pentagon Construction" by K. Knop
8. https://en.wikipedia.org/wiki/Henry_Martyn_Taylor. See also the web article "Henry Martyn Taylor, FRS" at http://trinitycollegechapel.com/about /memorials/brasses/taylor-hm/, and his obituary at https://academic.oup.com /mnras/article/89/4/324/1225534 (*Monthly Notices of the Royal Astronomical Society*, vol. 89, no. 4, February 1929).
9. See Kaplan, Robert, and Ellen Kaplan, *Hidden Harmonies: The Lives and Times of the Pythagorean Theorem* (New York: Bloomsbury Press, 2011), pp. 103–7. See also "The World of Blind Mathematicians" by Allyn Jackson, *Notices of the AMS*, vol. 49, no. 10, November 2002, p. 1247.

Pentagonal Numbers

PENTAGONAL NUMBERS ARE POSITIVE INTEGERS represented by equally spaced dots placed along the sides of a regular pentagon, as shown in figure S2.1. Start with a pentagon of one dot at each vertex and thus two dots on each side. Next, extend the top two sides so that each now contains three dots, and complete the structure to form a larger pentagon, whose remaining three sides are parallel to those of the original pentagon. The larger pentagon has seven new dots, so the total number of dots is now $5 + 7 = 12$. Now repeat the process with four dots to each side, producing ten new dots for a total of $12 + 10 = 22$. Continuing in this manner, we arrive at the sequence

$$1, 5, 12, 22, 35, 51, 70, 92, \ldots,$$

where the leading 1 stands for the "pentagon" consisting of just a single point. We note that the numbers follow the sequence O, O, E, E,..., where O stands for odd and E for even.

Can we predict the nth pentagonal number without calculating all the previous members of the sequence? Yes indeed. Let's look again at how the sequence is formed:

$$1 = 1, 5 = 1 + 4, 12 = 5 + 7, 22 = 12 + 10, 35 = 22 + 13, \ldots,$$

giving us the recursive formula

$$P_1 = 1, P_n = P_{n-1} + 3n - 2, n = 2, 3, 4, \ldots.$$

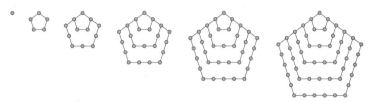

FIGURE S2.1. Pentagonal numbers

So we have

$$P_1 = 1, P_2 = 1 + 4 = 5, P_3 = 1 + 4 + 7 = 12,$$
$$P_4 = 1 + 4 + 7 + 10 = 22,$$

and generally $P_n = 1 + 4 + 7 + 10 + \cdots + (3n - 2)$. This is an arithmetic progression with the initial term 1 and common difference 3. To find its sum, we write the progression twice—once from the beginning to the end, and again backward, starting with the last term and ending with the first:

$$P_n = 1 + 4 + 7 + \cdots + (3n - 5) + (3n - 2)$$
$$P_n = (3n - 2) + (3n - 5) + \cdots + 7 + 4 + 1$$

Adding the terms vertically, each pair adds to $3n - 1$ and there are n such pairs, so we have $2P_n = n(3n - 1)$, and finally

$$P_n = \frac{n(3n - 1)}{2}.$$

For example, the tenth pentagonal number is 145, the hundredth number is 14,950, and so on.[1]

On rare occasions, it may happen that a pentagonal number is also a perfect square, represented by a square pattern of dots. The first five of these *pentagonal square numbers* are

$$P_1 = 1, P_{81} = 9{,}801 = 99^2, P_{7{,}921} = 94{,}109{,}401 = 9{,}701^2,$$
$$P_{776{,}161} = 903{,}638{,}458{,}801 = 950{,}599^2, \text{ and}$$
$$P_{76{,}055{,}841} = 8{,}676{,}736{,}387{,}298{,}001 = 93{,}149{,}001^2.$$

All known pentagonal square numbers end with 01, and their square roots end alternately with 1 and 9.[2]

NOTES AND SOURCES

1. The formula can also be verified by the process known as induction: we assume the formula is true for all $n = 1, 2, 3, \dots, m$ and then prove that it still holds true for $n = m + 1$. I leave the proof to the reader.

2. For additional properties of pentagonal square numbers, including a list of the first fifteen of them, see *The On-Line Encyclopedia of Integer Sequences* at https://oeis.org/A036353. See also the article "Pentagonal Square Numbers" at Wolfram MathWorld, https://mathworld.wolfram.com /PentagonalSquareNumber.html.

CHAPTER 5

The Pentagram

The triple interwoven triangle, the pentagram,
was used by the Pythagoreans as a symbol of
recognition between the members of the same
school, and was called by them *Health*.

**—SIR THOMAS HEATH, COMMENTARY
ON EUCLID'S *THE ELEMENTS* (1947)**

IF WE DRAW THE FIVE DIAGONALS of a regular pentagon,
we get a five-sided interlocking figure—a *pentagram* or
star pentagon. This peculiar shape has long been revered
as a symbol of virtue, fortune, and good luck. A bronze
coin from the ancient Greek town of Pitane and dated
around the fourth–third century BCE shows the head
of Zeus-Ammon in profile; the reverse side shows a pentagram with the Greek word *ΠΙΤΑΝ* inscribed around it
(figure 5.1). Pitane was located in the region of Mysia in
western Anatolia, near the modern town of Çandarli in
Turkey.[1] This is also the general region where Pythagoras spent his early years, presumably studying under
the great sage Thales of Miletus. Pythagoras may have
encountered the pentagram in his sojourns to Egypt and
Mesopotamia, where the five-cornered star was a common icon on city seals and tomb ornaments. His followers
adopted it as their emblem and are said to have used
it as a kind of ID that allowed members of the sect to
identify themselves to one another. According to a widely
accepted tradition, the pentagram's five corners represent

FIGURE 5.1. Ancient coin with a pentagram

the five elements of which the world is made—earth, water, air, fire, and the ethereal spirit that hovers above them all.

The pentagram soon became a recognizable icon throughout the Hellenistic world, where it was thought to be capable of fending off evil spirits. In the kingdom of Judea, it was stamped on pottery vessels, perhaps as proof of purchase for the tax authority. As reported by Oded Lipschits and Efrat Botcher of Tel Aviv University, embossed in each of the pentagram's five recesses was one letter of the Hebrew word ירשלם , YRŠLM, "Jerusalem" (Hebrew words are generally written without vowels, their pronunciation being taught to children at a young age). Figure 5.2 shows an example dated ca. 400–200 BCE.[2]

In the Muslim world, the pentagram is often known as Solomon's Seal (although this name is also used for the hexagram, or Star of David). According to one tradition, its five points stand for love, truth, peace, freedom, and justice. A beautifully preserved pentagram decorates the masonry surrounding the Jaffa Gate in Jerusalem, the western entrance to the Old City (figure 5.3). It was carved in stone by Ottoman ruler Suleiman the Magnificent

FIGURE 5.2. A pottery seal from ancient Judea

FIGURE 5.3. Decorative pentagram at the Jaffa Gate, Jerusalem

in 1541 upon completion of his grand project, building the massive walls around the city. A pentagram—interlocked in some versions—appears on the national flag of Morocco, as well as on the flag of its region of Tangier (see plate 8).[3]

Associations with the pentagram, however, are not always positive. To some, it is a sign of witchcraft or evil. On December 13, 2012, the *Chicago Tribune* reported the following:

> Houston—A North Texas man carved a pentagram into his 6-year-old son's back, telling a police dispatcher that he did it because 12-12-12 is a "holy day." ... The police said the boy, who was in stable condition, apparently was assaulted with a box cutter that officers found in the house.

For some, the pentagram symbolizes the magic power of wizardry. In his drama *Faust I*, Johann Wolfgang von Goethe alludes to it in this dialogue:[4]

> Mephistopheles: Let me admit; a tiny obstacle
>   Forbids me walking out of here:
>   It is the druid's foot upon your threshold.
> Faust: The pentagram distresses you?
>   But tell me, then, you son of hell.
>   If this impedes you, how did you come in?
>   How can your kind of spirit be deceived?
> Mephistopheles: Observe! The lines are poorly drawn;
>   That one, the angle pointing upward,
>   Is, you see, a little open.

★　◆　★

The pentagon and its associated pentagram can be drawn with a single stroke of the pen, tracing each side just once (this can be extended to include the circumscribing circle; see figure 5.4). In a sense, the two shapes comprise the two components of a single geometric structure. In fact,

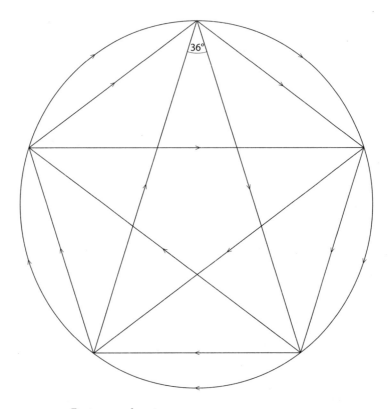

FIGURE 5.4. Pentagon and pentagram

we can just as well think of the *pentagram* as the defining shape, from which we obtain a pentagon by joining pairs of adjacent corners with straight lines. These lines can be extended outward from the ensuing pentagon, where their points of intersection form a larger, inverted pentagon—a process known as *stellation* (figure 5.5). The process can be repeated in either direction: starting with a given pentagon, join pairs of nonadjacent corners, and you get a smaller, inner pentagon; joining pairs of nonadjacent *lines*, you get a larger outer pentagon, and so on ad infinitum. Plate 9, *Pentagons and Pentagrams* by Eugen Jost, shows this process carried out to four levels,

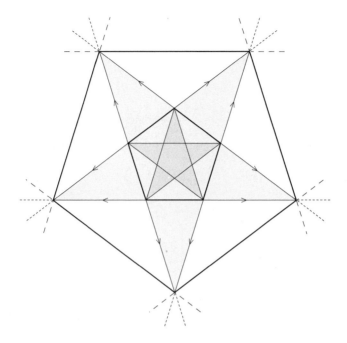

FIGURE 5.5. Stellated pentagon

with the pentagons shown in orange and the pentagrams in blue.

This dual relationship gives rise to several interesting features, all of which involve the golden ratio. To begin with, note that two adjacent sides of a pentagram, together with the side opposite their common vertex, form a golden triangle—an isosceles triangle with a top angle of 36 degrees. This means that the side-to-base ratio of this triangle is φ, the golden ratio (see figure 4.7). And since the pentagram comprises five of these sides, its total length is 5φ (taking the base to be of length 1).

The pentagram's five golden triangles enclose a smaller pentagon at the center, and we now find its dimensions. Denote the length of its side by x and the length of each of the two "arms" on its sides by y (figure 5.6). We have

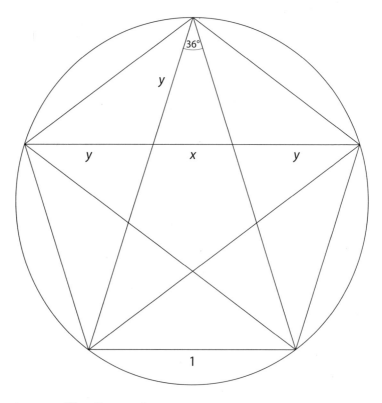

FIGURE 5.6. Dissecting a pentagram

$x + 2y = \varphi$ (because the horizontal line is also a diagonal). Now the small upper triangle is similar to the larger triangle in which it is embedded, so we have $x/1 = y/\varphi$. Solving these two equations and using the relation $\varphi^3 = 1 + 2\varphi$, we get

$$x = 1/\varphi^2, \quad y = 1/\varphi.$$

Thus, the inner pentagon is a smaller, inverted replica of the outer pentagon and of side $1/\varphi^2 \sim 0.382$ of the original side. Consequently, its *area* is $1/\varphi^4 \sim 0.146$, or about 15 percent of the area of its progenitor.

We can now repeat the process and create smaller and smaller pentagons and pentagrams, each nested inside its predecessor, ad infinitum. These pentagons have sides of length $1/\varphi^2$, $1/\varphi^4$, $1/\varphi^6$,..., of the side of the initial pentagon. Out of curiosity, I tried to find the total sum of the perimeters of these pentagons, beginning with a pentagon of unit side:

$$5(1 + 1/\varphi^2 + 1/\varphi^4 + 1/\varphi^6 + \cdots).$$

The expression inside the parentheses is an infinite geometric progression, or series, with the common ratio $r = 1/\varphi^2$. Such a series converges to the sum $S = 1/(1-r)$, provided that $-1 < r < 1$. And since $1/\varphi^2 \sim 0.382 < 1$, the series converges to the limit $S = \dfrac{1}{1 - 1/\varphi^2} = \dfrac{\varphi^2}{\varphi^2 - 1} = \dfrac{\varphi^2}{\varphi} = \varphi$. So the total length of all these nested pentagons is 5φ. But this is also the length of the *pentagram* inscribed in the original pentagon. We could therefore construct an Alexander Calder–style mobile, where all these pentagons, each represented by its perimeter, would be balanced by a single pentagram. Well, at least in theory; providing for an infinite number of pentagonal frames would be beyond human capacity.

All this shows that the golden ratio is central to an understanding of the pentagon-pentagram system. In fact, φ is to the regular pentagon what π is to the circle or e to the logarithmic spiral. Moreover, it gives us two ways of writing the various formulas associated with the pentagon: we can express them in terms of φ (as a single quantity) or in terms of $\sqrt{5}$, whichever is more convenient.[5]

Besides its close relation to the pentagon, the pentagram has the property that each of its five sides divides any other side in the golden ratio. To see this, we refer again to figure 5.6. Each of the two diagonals through the top vertex divides the horizontal diagonal into two segments, $y + x$ and y. Their ratio is

$$\frac{y + x}{y} = \frac{1/\varphi + 1/\varphi^2}{1/\varphi} = 1 + \frac{1}{\varphi} = \varphi,$$

which is also the ratio of the whole diagonal $2y + x$ to the long part $y + x$, as the reader can easily confirm (see appendix D, figure D.1 for a summary of these relations). This discovery is credited to Theatetus of Athens (ca. 414–369 BCE), a close friend of Plato and one of the very few Greek philosophers of whom we have some details about their lives and personalities. Plato paid tribute to him in his *Dialogues*, where he described Theatetus as having "a turned-up nose and slightly protruding eyes, an uncommonly penetrating mind, was very quick in his thinking, had a remarkable memory, and with all this possessed a gentleness of character which seems to have been the distinguishing mark of his personality."[6] His main contribution to mathematics was the study of irrational numbers (or in the Greek terminology, incommensurable quantities), which were still somewhat of a mystery at his time. He made a distinction between irrational numbers such as $\sqrt{5}$, whose square is rational, and numbers like the golden ratio $(1 + \sqrt{5})/2$, whose square $(3 + \sqrt{5})/2$ is still irrational. Today, of course, we do not make such a distinction.

★ ● ★

Earlier in this chapter, we saw that the area of the pentagon enclosed inside a pentagram is $1/\varphi^4 \sim 0.146$ that of its progenitor. But we still need to find the actual area of the original pentagon, taking its side to be 1. To this end, we can follow two approaches. The easy way is to divide the

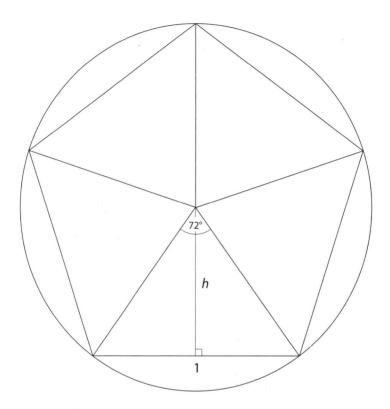

FIGURE 5.7. Dissecting a pentagon I

pentagon into five identical sectors, each an isosceles triangle with base = 1 and top angle = 72° (figure 5.7). Using trigonometry, the height of each triangle is $h = \frac{1}{2}\cot 36°$ and its area = (base × height)/2 = $\frac{1}{4}\cot 36° \sim 0.344$. The total area of the pentagon is five times this value, or about 1.720.

Yes, this is the easy approach, but not exactly the "proper" one. It violates an unwritten rule among classical geometers that all constructions—as well as any calculations based on them—should be done with Euclidean tools alone and rely only on proved geometric theorems, preferably those included in Euclid's *Elements*. This at once

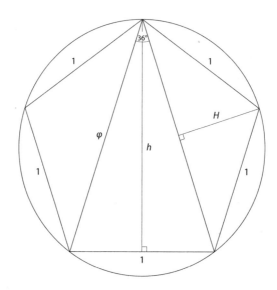

FIGURE 5.8. Dissecting a pentagon II

excludes trigonometry from the game. So let us follow the "pure" approach. We divide the pentagon into three triangles—a golden triangle with base 1 and sides φ, flanked by two isosceles triangles with base φ and sides 1 (figure 5.8).

To find the areas of these triangles, we need to find their heights. Denoting the height of the central triangle by h, we have $(1/2)^2 + h^2 = \varphi^2$, so $h^2 = \varphi^2 - 1/4 = (1+\varphi) - 1/4 = (3+4\varphi)/4$ and $h = \sqrt{3+4\varphi}/2$. This gives us the area of the middle triangle as $(1 \times h)/2 = \sqrt{3+4\varphi}/4$.

Moving on to the two outer triangles and denoting their heights by H, we have $(\varphi/2)^2 + H^2 = 1^2$, so $H^2 = 1 - \varphi^2/4 = \dfrac{3-\varphi}{4}$ (where we again replaced φ^2 by $1+\varphi$) and $H = \sqrt{3-\varphi}/2$. This gives us the area of each of the outer triangles as $(\varphi \times H)/2 = \varphi\sqrt{3-\varphi}/4$. But now we do a little manipulation to demonstrate the advantage of using the single symbol φ instead of its explicit expression

$(1 + \sqrt{5})/2$: we move the φ in front of the radical sign to the inside, changing it to $\varphi^2 = 1 + \varphi$. This gives us the area as $\sqrt{(1 + \varphi)(3 - \varphi)}/4 = \sqrt{3 + 2\varphi - \varphi^2}/4$, and replacing φ^2 yet again by $1 + \varphi$, we get $\sqrt{2 + \varphi}/4$. The combined area of the two side triangles is thus $\sqrt{2 + \varphi}/2$, and adding this to the area of the middle triangle, we finally get the result:

Area of pentagon of unit side $= \sqrt{3 + 4\varphi}/4 + \sqrt{2 + \varphi}/2$.

Now this was not exactly an inspiring process, and the net result is even less appealing—a sum of two radicals that seemingly cannot be combined into a single expression. But it can! So as not to spoil the fun, I give here only the result and leave the proof—which is not particularly difficult but involves even more algebraic manipulations with powers of φ—to the reader:

$$\text{Area of pentagon of unit side} = \frac{\sqrt{5(3 + 4\varphi)}}{4}.$$

You notice that at no point in this derivation did we ever use the actual value of φ, that is, $(1 + \sqrt{5})/2$. But since geometry books usually give the result in the latter form, we give it here:

$$\text{Area of pentagon of unit side} = \frac{\sqrt{5(5 + 2\sqrt{5})}}{4}.$$

We see here the advantage of using φ as the basic numerical "building block" of the pentagon-pentagram system; it is the natural choice for all things pentagonal.

NOTES AND SOURCES

1. Adapted from the website Ancient to Medieval (and Slightly Later) History at https://archaicwonder.tumblr.com/post/76908550286/rare-bronze-coin-from -pitane-mysia-c-4th-3rd.

2. Oded Lipschits and Efrat Botcher, "The YRŠLM Stamp Impressions on Jar Handles: Distribution, Chronology, Iconography and Function," *Tel Aviv*, vol. 40, 2013, p. 107. See also "Strata: Pentagrams in Judea," Biblical Archeology Society Online Archive, November/December 2013, at https://www .baslibrary.org/biblical-archaeology-review/39/6/18.
3. "List of Moroccan Flags" at https://en.wikipedia.org/wiki/List_of_Moroccan _flags.

    More articles and images of pentagrams can be found at websites dealing with occult, spiritual, and religious symbolism. Here is a small sample:
    - ● "A Brief History of the Pentagram" at https://willyoctora.wordpress .com/2013/07/08/a-brief-history-of-the-pentagram/
    - ● "Pentagram" at https://en.wikipedia.org/wiki/Pentagram
    - ● "The Pentagram through History" by Lionel Pepper at https://www.spiritualdimensions.co.nz/spiritual-learning/wicca/the -pentagram/the-pentagram-through-history/
    - ● "Symbolic Meaning of the Pentagram" at http://www.globalstone.de /essays/The_history_of_pentagram.pdf
    - ● "Pentagram" at https://slife.org/pentagram/
4. Gyorgy Kepes, ed., *Module, Symmetry, Proportion*, p. 191.
5. Indeed, one wishes that scientific calculators would have a special key for φ just as they have for π and e.
6. François Lasserre, *The Birth of Mathematics in the Age of Plato* (Larchmont, NY: American Research Council, 1964), pp. 65–66. The quotation is in Lasserre's words.

PLATE 1. Plumeria

PLATE 2. Petunia

PLATE 3. Stained glass window at the Stephansmünster in Breisach, Germany

PLATE 4. Eugen Jost, *Hamsah*

PLATE 5. Charles Demuth, *I Saw the Figure 5 in Gold*

PLATE 6. Eugen Jost, *All Is Five*

$$17 = 2^4 + 1$$

PLATE 7. Eugen Jost, *Homage to Gauss*

PLATE 8. Flag of Tangier, Morocco

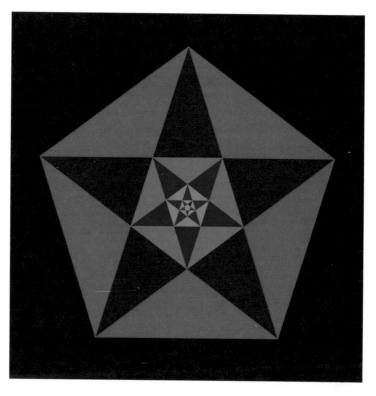

PLATE 9. Eugen Jost, *Pentagons and Pentagrams*

PLATE 10. *Marocaster coronatus*

PLATE 11. The Great Star Flag

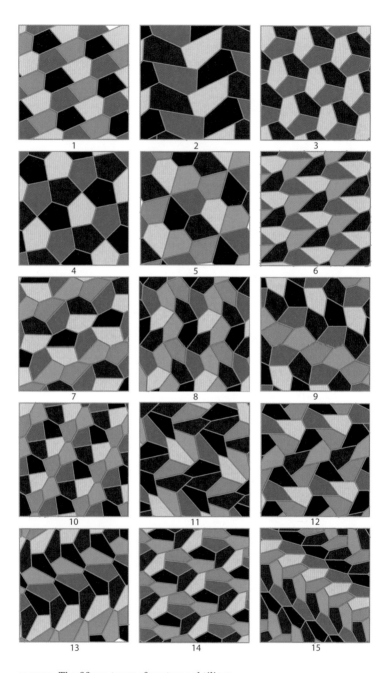

PLATE 12. The fifteen types of pentagonal tilings

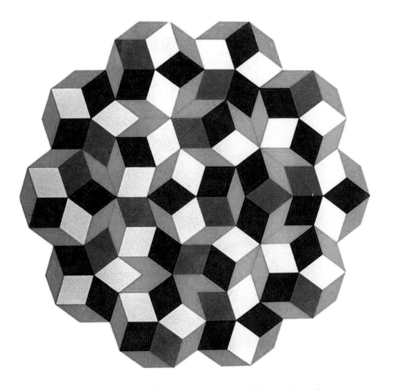

PLATE 13. Penrose tiling at the Weizmann Institute, Rehovot, Israel

PLATE 14. Eugen Jost, *Kites and Darts*

PLATE 15. Decorative design at the Bibi-Khanym mosque, Samarkand, Uzbekistan

PLATE 16. Israeli postage stamp commemorating
the discovery of quasiperiodic crystals

PLATE 17. Citadel of Jaca, Huesca, Spain

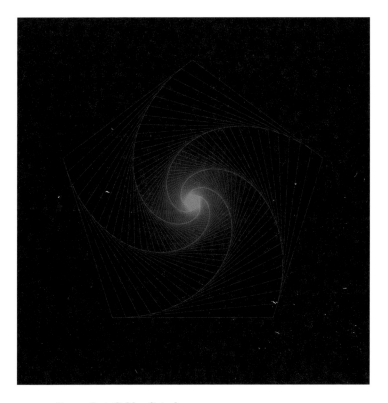

PLATE 18. Eugen Jost, *Golden Spirals*

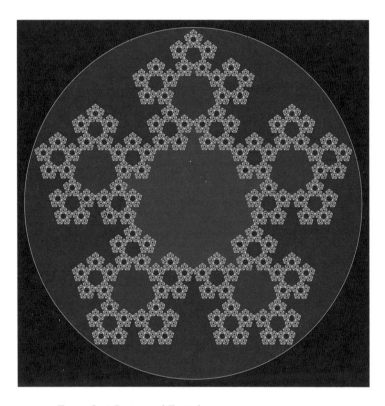

PLATE 19. Eugen Jost, *Pentagonal Fractals*

CHAPTER 6

Pentastars

Twinkle, twinkle little star
How I wonder what you are!

—JANE TAYLOR (1806)

IF WE SHADE THE AREA OF A PENTAGRAM, we get what I
call a *pentastar*—a five-pointed star and one of the most
frequently found symbols anywhere. Pentastars appear
on the national flags of 48 of the world's 185 countries
(plus the European Union), whereas only three display
a hexastar (Burundi, Equatorial Guinea, and Slovenia)
and one shows a hexagram, or Star of David (Israel). The
pentastar is perhaps the most visible Christmas decoration
worldwide—probably as homage to the mythological Star
of Bethlehem, depicted in numerous Christian images.
(Ironically, when you view a star in a slightly out-of-focus
telescope, you often see a six-spiked diffraction pattern, not
a pentastar.) The allure of the pentagram, from which this
figure derives, seems irresistible.

The name *pentastar* is not an official word; it was coined
by the Chrysler Corporation as a brand name in 1962, when
the company was in the process of creating a new "Corporate Identity Office concerned with the manner in which
the company identifies and visually presents itself and
its products to the public."[1] Chrysler received eight hundred proposed designs; it chose the pentastar. Explained
Robert Stanley, who created the design and gave it its
name, "We were looking for something that would not be

FIGURE 6.1. Chrysler's Pentastar logo

too complicated for people to remember and still have a very strong, engineering look to it … something simple, a classic, geometric form that was not stolid. That's why we broke up the pentagonal form that became the Pentastar. It provided a certain tension and a dynamic quality." The Pentastar was embossed onto the front cover of the company's 1962 annual report (figure 6.1). The iconic corporate symbol remained in use until the company was taken over by Daimler in 1998, when it was removed.

An entirely different pentastar design is at the heart of an unusual map of the world, on which each of the five major continents appears on one petal of a pentastar (figure 6.2). It was presented in 1879 by German geographer Heinrich Karl Wilhelm Berghaus (1797–1884) and is known as the Berghaus star projection. The map is centered on the North Pole and shows all meridians (circles of longitude) as straight lines, broken at the equator (except for the central meridian of each petal); they emanate from the center and terminate close to the South Pole. The parallels (circles of latitude) are concentric circles around the North Pole. The Northern Hemisphere uses the azimuthal equidistant projection, in which every point on the globe is represented in proportion to its true distance from the North Pole.

Berghaus's map is rarely used in practice; it is known chiefly for its neat appearance, as it divides the world in a rather convenient way. But real continents are not arranged on the globe in a precise geometric shape, so it is inevitable that some portions of each continent spill over

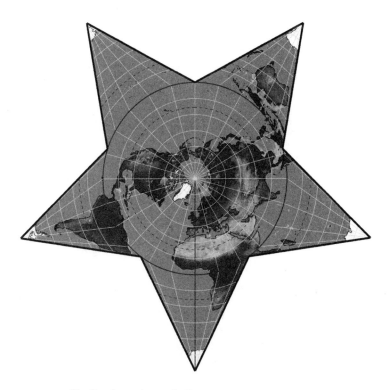

FIGURE 6.2. The Berghaus star projection

to an adjacent petal. In 1904 Berghaus's map earned a
place of honor as the logo of the American Association of
Geographers (figure 6.3).

The five continents that Berghaus grouped so neatly
in his map were not always arranged so conveniently;
according to German geophysicist, meteorologist, and arctic explorer Alfred Lothar Wegener (1880–1930), Earth's
major landmasses were once lumped together in a single,
gigantic landmass, called Pangaea. It started to break apart
some 175 million years ago in a process he called "conti-
nental drift," known today as plate tectonics. At that
time, the shapes of the resulting landmasses were vastly
different from their modern appearance. Yet, against

FIGURE 6.3. Logo of the American Association of Geographers

all odds, some marine creatures survived this colossal reshaping of Earth's surface unscathed. Among them is a fossilized starfish known as *Marocaster coronatus*, found near the city of Marrakesh in Morocco and preserved in pristine condition (it is now at the Muséum de Toulouse in France; see plate 10). It was dated to the geologic Cretaceous period, roughly 145 to 66 million years ago. Its perfect symmetry and minute details that have been frozen in time over countless millennia cannot but leave you in awe at the workings of nature; I like to think of it as a greeting card left to us unchanged since the day it was imprinted in that rock long before humans even existed on this planet.

★　●　★

The flag of the United States of America has undergone numerous variations in the country's nearly 250-year history. Every time a new state joined the Union, another pentastar was added to the flag. As a result, the pattern in which these stars were arranged had to be frequently changed, resulting in some interesting geometric designs. One of the least known of these variations is the Great Star Flag, comprising twenty pentastars placed along the sides of a large pentagram (see plate 11). It was the official flag of the United States for just one year, from July 4, 1818, to July 3, 1819, although it flew over the Capitol dome for half a year prior. Its design is credited to US Navy captain Samuel Chester Reid (1783–1861), who suggested

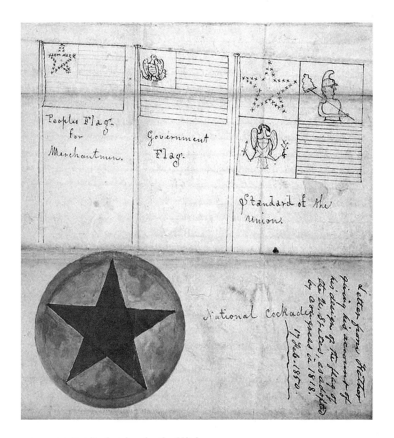

FIGURE 6.4. Reid's sketches for the US flag

that the flag should have thirteen stripes in honor of the thirteen founding states of the Union, and twenty pentastars for the twenty states comprising the Union in 1818. His design was adopted by Congress and signed into law as the Flag Act of 1818 by President James Monroe on April 4 of that year. Figure 6.4 shows Reid's sketches for his design, from an 1850 letter to his son. The design was later changed to a rectangular array of four rows with five stars in each.[2]

★ ⬠ ★

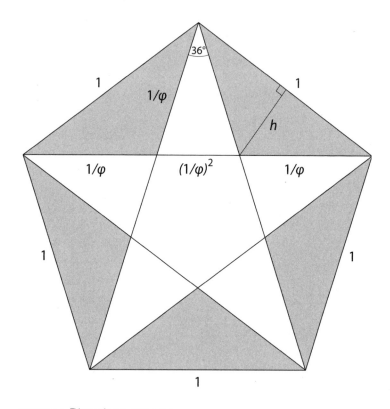

FIGURE 6.5. Dissecting a pentastar

The iconic Chrysler logo may have been "something simple, a classic, geometric form that was not stolid" to its creator, Robert Stanley, but as we have seen in previous chapters, the pentagon and pentagram are anything but simple when attempting to find their areas. The same goes for the perimeter and area of the pentastar.

To find the pentastar's total length, or perimeter, we connect its five vertices to form a pentagon of unit side (figure 6.5). As we saw in chapter 5, the length of each of the pentastar's "arms" is $1/\varphi$. Its perimeter is therefore $10 \cdot (1/\varphi) = 10 \cdot (\sqrt{5} - 1)/2 \sim 6.180$, or about 1.236 times the perimeter of the generating pentagon.

To find its area, we denote the height of each of the shaded triangles surrounding the pentastar on the outside by h. We have $h^2 + (1/2)^2 = (1/\varphi)^2 = 2 - \varphi$, so $h^2 = (2 - \varphi) - 1/4 = (7 - 4\varphi)/4$ and $h = \sqrt{7 - 4\varphi}/2$. The area of each shaded triangle is therefore $(1/2) \cdot 1 \cdot h = \sqrt{7 - 4\varphi}/4$. We now subtract five of these areas from the area of the outside pentagon, which we found in chapter 5 to be $\dfrac{\sqrt{5(3 + 4\varphi)}}{4}$:

$$\text{Area of pentastar} = \frac{\sqrt{5(3 + 4\varphi)}}{4} - \frac{5}{4}\sqrt{(7 - 4\varphi)}.$$

This ungainly expression can be manipulated into something simpler, using again the linearization formulas for powers of φ. The net result is $\dfrac{5}{2\sqrt{3 + 4\varphi}}$. When this is transformed into actual numbers, we finally get the required area:

$$\text{Area of pentastar} = \frac{5}{2\sqrt{5 + 2\sqrt{5}}} = \frac{\sqrt{5(5 - 2\sqrt{5})}}{2} \sim 0.812.$$

This amounts to about 47 percent of the area of the generating pentagon.

As we said at the beginning of this book, nothing about the pentagon is as simple as meets the eye.

NOTES AND SOURCES

1. The quotes in this paragraph are from the web article "Chrysler's Pentastar" at https://www.allpar.com/corporate/pentastar.html.
2. The information about the Great Star Flag was culled from four web articles: "Samuel Chester Reid" at https://en.wikipedia.org/wiki/Samuel_Chester _Reid; "Flag of the United States" at https://en.wikipedia.org/wiki/Flag_of _the_United_States; "Rare Flags" at http://www.rareflags.com?RareFlags _Collecting_GreatStar.htm; and "20 Star Flag—(1818–1819) (U.S.)" at https:// www.crwflags.com/fotw/flags/us-1818.html.

Pentagonal Puzzles
and Curiosa

HENRY ERNEST DUDENEY (1857–1930) has long been regarded as England's greatest creator of mathematical puzzles. He was to nineteenth- and early twentieth-century readers what Martin Gardner (1914–2010), in his numerous books and monthly column in *Scientific American*, was to his twentieth- and early twenty-first-century followers. Like Gardner, Dudeney never had formal mathematical training, yet he had a gift for solving difficult problems that often defied conventional thinking. He wrote six books on recreational mathematics; they are delightful works, not only for the hundreds of puzzles they offer but also for the charming, hand-drawn illustrations that accompany many of them. The earliest of his books, *The Canterbury Puzzles*,[1] is based on characters from Chaucer's *The Canterbury Tales*. In this book, we find the following puzzles, quoted here verbatim (solutions can be found in appendix E).

18. THE SHIPMAN'S PUZZLE.

Of this person we are told, "He knew well all the havens, as they were, From Gothland to the Cape of Finisterre, And every creek in Brittany and Spain: His barque yclepéd was the *Magdalen*." The strange puzzle in navigation that he propounded was as follows:

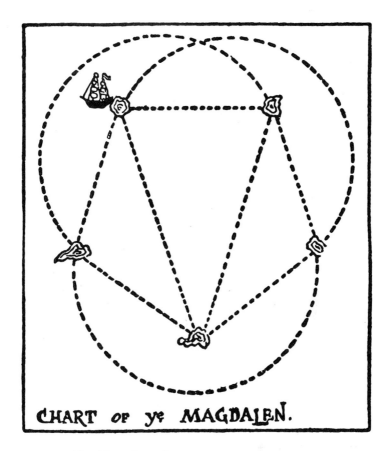

CHART OF ye MAGDALEN.

FIGURE S3.1. The shipman's puzzle

"Here be a chart," quoth the Shipman, "of five islands, with the inhabitants of which I do trade [see figure S3.1]. In each year my good ship doth sail over every one of the ten courses depicted thereon, but never may she pass along the same course twice in any year. Is there any among the company who can tell me in how many different ways I may direct the *Magdalen's* ten yearly voyages always setting out from the same island?"

21. THE PLOUGHMAN'S PUZZLE.

The Ploughman—of whom Chaucer remarked, "A worker true and very good was he, Living in perfect peace and charity"—protested that riddles were not for simple minds like his, but he would show the good pilgrims, if they willed it, one that he had frequently heard certain clever folk in his own neighbourhood discuss.

"The lord of the manor in the part of Sussex whence I come hath a plantation of sixteen fair oak trees, and they be so set out that they make twelve rows with four trees in every row. Once on a time a man of deep learning, who happened to be travelling in those parts, did say that the sixteen trees might have been so planted that they would make so many as fifteen straight rows, with four trees in every row thereof. Can ye show me how this might be? Many have doubted that 'twere possible to be done." The illustration [see figure S3.2] shows one of many ways of forming the twelve rows. How can we make fifteen?

Can *you* do it?

The following puzzle is taken from Dudeney's second book, *Amusements in Mathematics*:[2]

155. THE PENTAGON AND SQUARE.

I wonder how many of my readers, amongst those who have not given any close attention to the elements of geometry, could draw a regular pentagon, or five-sided figure, if they suddenly required to do so. A regular hexagon, or six-sided figure, is easy enough…But a pentagon is quite another matter. So, as my puzzle has to do with

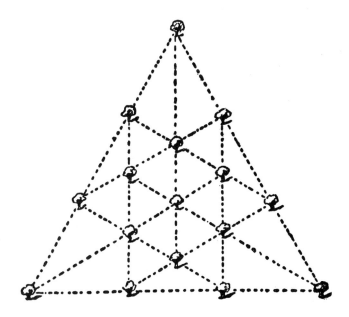

FIGURE S3.2. The ploughman's puzzle: sixteen trees, twelve rows, four trees per row

the cutting up of a regular pentagon, it will perhaps be well if I first show my less experienced readers how this figure is to be correctly drawn. [The author then outlines the construction we discussed on page 50.]

Having formed your pentagon, the puzzle is to cut it into the fewest possible pieces that will fit together and form a perfect square.

Please be warned: this is not an easy puzzle. You can find the solution in appendix E.

★ ⬟ ★

So much for Dudeney's puzzles. A much easier brainteaser is this: How many distinct triangles are there in figure S3.3?[3] See appendix E for the solution.

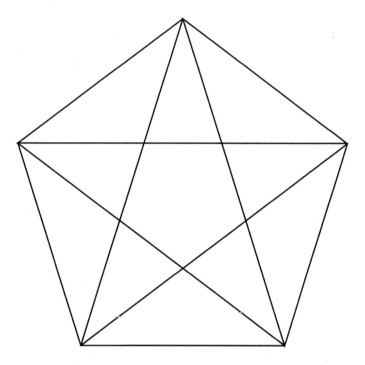

FIGURE S3.3. How many triangles?

An old Chinese puzzle, the *tangram*, consists of a flat wooden square partitioned into seven pieces of different shapes. The goal is to disassemble these pieces, scramble them, and reassemble them to get the original square. As a side product, hundreds of shapes, from human figures to animals and inanimate objects, can be constructed from any combination of the pieces; the number of possibilities is limited only by one's imagination. Figure S3.4 shows a simplified version of a tangram, consisting of five scrambled pieces. Can you reassemble them into a regular pentagon? The design is by Eugen Jost. You'll find the solution in appendix E.

FIGURE S3.4. A scrambled pentagonal tangram

And then there are pentagonal mazes. Not many outdoor examples of them exist, like the famed maze gardens of England and France—in fact, almost none. But quite a few graphic artists have created intriguing pentagonal mazes on paper or on computer screens. Figure S3.5 shows one example.

Figure S3.6 is an Escher-style print showing a nest of five snakes, each emerging from a different dimension to meet the others at the center. The design is by Belgian artist Peter Raedschelders. In his own words,

FIGURE S3.5. Pentagonal maze

The combination of art and mathematics is my hobby; I am neither artist nor mathematician. The artworks I make are prints made from hand-made drawings. I have been highly influenced by the work of M. C. Escher, but I try to find mathematical ideas that Escher never used. Most of my works are tilings…, but some are prints with an unusual perceptive.[4]

★ ◈ ★

Imagine five bugs located at the corners of a regular pentagon. At the stroke of a signal, each bug starts to move toward its neighbor. What path will they follow, and where

FIGURE S3.6. Peter Raedschelders, *Five Snakes*

will they meet? The paths turn out to be logarithmic spirals that converge at the pentagon's centroid. Plate 18, *Golden Spirals* by Eugen Jost, shows how the spirals are formed entirely by their tangent lines, demonstrating that a curve can be generated not only by the set of points on it but also by the set of lines tangent to it.

Another design by Jost, *Pentagonal Fractals*, shows a self-similar pattern of ever smaller pentagons and pentagrams, all clustered around a central pentagonal void (plate 19). No matter how much you enlarge it, every detail in the pattern looks exactly like its smaller replica, repeating itself ad infinitum.

★ ⬠ ★

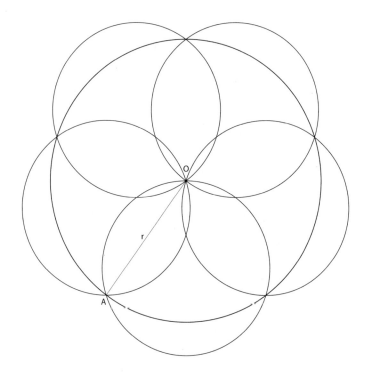

FIGURE S3.7. The five-disc problem

Here is a puzzle that is really a geometry problem in its own right: Five equal discs of unit radius are placed symmetrically as shown in figure S3.7, so that their centers form the corners of a regular pentagon and their circumferences all pass through the pentagon's centroid O. What is the radius of the largest circular area covered by the five discs (i.e., what is the length of OA)?[5] See appendix E for the solution.

NOTES AND SOURCES

1. H. E. Dudeney, *The Canterbury Puzzles* (New York: Dover, 1958). Originally published in 1907.
2. H. E. Dudeney, *Amusements in Mathematics* (New York: Dover, 1958), pp. 37–38. Originally published in 1917.

3. Boris A. Kordemsky, *The Moscow Puzzles* (New York: Charles Scribner's Sons, 1972), pp. 2, 186.

4. Quoted from https://www.leonardo.info/gallery/gallery331/raedschelders .html.

5. Quoted from H. E. Huntley, *The Divine Proportion: A Study in Mathematical Beauty*, p. 45.

CHAPTER 7

Tessellations

DEFYING THE IMPOSSIBLE

"There's no use trying," she said:
"one can't believe impossible things."

—LEWIS CARROLL (CHARLES LUTWIDGE DODGSON),
THE HUNTING OF THE SNARK (1876)

THE ART OF TESSELLATION—filling the plane with identical shapes without leaving gaps or overlaps—has occupied the minds of artists and artisans since ancient times. You only need to witness the exquisite geometric tilings typical of Islamic architecture to realize how central this kind of decoration was to Islam. But carrying it out proved to be not as easy as it sounds; indeed, the subject falls squarely into several areas of mathematics, including geometry, elementary and advanced algebra, and in particular group theory. This is because any attempt to cover a flat surface with identical shapes must face the constraints imposed on it by the laws of geometry.

The simplest kind of tessellation involves regular polygons, but only a small minority of them will do: an equilateral triangle, a square, and a hexagon (figure 7.1).[1]

Glaringly missing from this short list is the regular pentagon. A quick look at figure 7.2 shows us why: any two of the pentagon's adjacent sides form an angle of 108 degrees between them, so for any number of regular pentagons to join

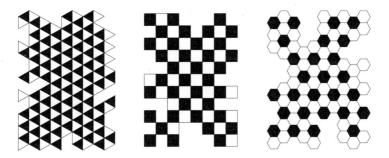

FIGURE 7.1. The three possible tessellations by regular polygons

at a vertex, 108 would have to go an integral number of times into a full rotation, that is, into 360 degrees—which is not the case. Similarly, all other *n*-gons—regular polygons with *n* sides—can be eliminated, except for $n = 3$, 4, and 6.

But what if we relax the restriction that the pentagon must be regular? This possibility has intrigued artists, artisans, and mathematicians for centuries. That such tilings do exist can be seen by the example shown in figure 7.3: its basic building block is reminiscent of the typical house that young children love to draw—a square with its top side replaced by a "roof" of two equal segments meeting at 90 degrees. The resulting pentagon can neatly tile the plane, and in fact in more than one way; figure 7.3 shows two such tilings.

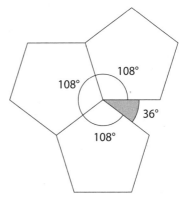

FIGURE 7.2. A regular pentagon cannot tile the plane

★ ⬟ ★

The systematic search to determine what kinds of pentagonal shapes can tile the plane began with Karl August

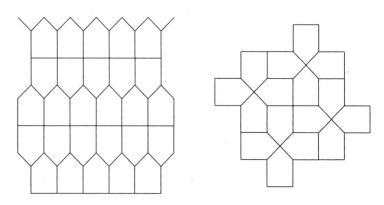

FIGURE 7.3. Two tilings with nonregular pentagons

Reinhardt (1895–1941), a German mathematician who was for a while assistant to David Hilbert, Europe's foremost mathematician at the time. Reinhardt became interested in Hilbert's eighteenth problem, which deals with tilings in two- and three-dimensional spaces. (The designation "eighteenth" refers to Hilbert's list of twenty-three problems that he considered to be the most important unsolved mathematical problems at the time; he presented them in 1900 at the Second International Congress of Mathematicians in Paris.) In 1918 Reinhardt discovered five types of congruent convex pentagons that tile the plane (the designation "convex" means that all interior vertex angles are less than 180°). He characterized each type by the relations between the various angles and sides of the pentagon. For example, type 1 is marked by the relations $\beta + \gamma = 180°$, $\alpha + \delta + \varepsilon = 360°$, with the sides having arbitrary length, while type 2 is characterized by $\beta + \delta = 180°$, $c = e$ (here and elsewhere, the vertex angles $\alpha, \beta, \gamma, \delta, \varepsilon$ are counted clockwise, and the sides a, b, c, d, e are labeled after the vertices from which they start, again in a clockwise sense). Figure 7.4 shows three examples of type 1 tilings. We note that in some cases it is not individual pentagons that tile the plane,

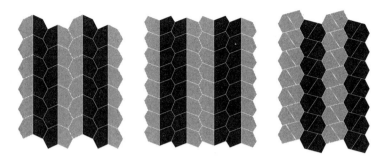

FIGURE 7.4. Three examples of type 1 tilings

but groups of congruent pentagons combined in various arrangements.[2]

Things thus stood for fifty years until 1968, when Richard B. Kershner of Johns Hopkins University discovered three more types of convex pentagons that could tessellate the plane. Kershner also proved that it is impossible to tessellate the plane by convex n-gons for $n > 6$; that is, only triangles, quadrilaterals, pentagons, and hexagons can tile the plane, and no other convex polygons. In his article announcing the discovery, Kershner also claimed that this list exhausts all possible tilings by congruent convex pentagons. However, he omitted his proof, saying only that it is "extremely laborious."[3]

★ ⬠ ★

In July 1975 the journal *Scientific American* ran an article on tessellations by the journal's legendary mathematics columnist, Martin Gardner. It instantly popularized the subject and generated a great deal of interest among professionals and laypeople alike. Coincidentally, in December of that year, Richard James, a computer programmer, discovered a new type of convex pentagonal tiling, raising the total number to nine.

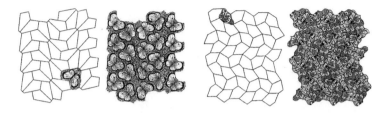

FIGURE 7.5. Two of Rice's pentagonal tilings rendered as works of art

Now enter Marjorie Rice (1923–2017), a San Diego housewife raising five children and having no formal mathematical training beyond a one-year high school math class. She read Gardner's article and started to play with pentagonal shapes, using her kitchen countertop as a playground for shapes and colors. To her own surprise, she discovered four new pentagonal tessellations that no one had known about; but lacking confidence in their veracity, she sent them to Gardner, who in turn sent Rice's tilings to Doris Schattschneider, a mathematics professor at Moravian College in Pennsylvania and an acknowledged expert on tessellations. Schattschneider confirmed Rice's findings, but not before she had to familiarize herself with Rice's self-invented notation.

But Rice was not content with just creating bland, abstract pentagonal tiles; inspired by the all-time master of tessellations, Maurits Cornelis Escher, she wanted to give them life in the form of flowers, butterflies, and bees, thereby transforming her geometric discoveries into exquisite works of art (figure 7.5). To commemorate Rice's achievement, in 1999 the foyer floor at the offices of the Mathematical Association of America in Washington, DC, was decorated with a glazed tiling of one of the four types of tessellations she discovered (figure 7.6).

Rice lived to the age of ninety-four; although her final years saw her health in decline, she lived to witness the discovery of two more pentagonal tilings, one by Rolf Stein

FIGURE 7.6. Mathematical Association of America foyer floor tiling in honor of Marjorie Rice

in 1985 and another by Casey Mann, Jennifer McLoud-Mann, and David Von Derau of the University of Washington Bothell in 2015. This raised the number of known pentagonal tessellations to fifteen. Then on August 1, 2017—exactly one month after Rice's death—Michaël Rao (b. 1980), a mathematician at CNRS (France's national center for scientific research) and the École Normale Supérieure de Lyon, posted the results of his exhaustive computer search, which showed that the list is complete at fifteen.[4] Thus, the long quest to find all possible pentagonal convex shapes that tessellate the plane achieved its goal, almost exactly one hundred years after Reinhardt had initiated it in 1918. Examples of all fifteen types are shown in plate 12, while their characteristics are listed in figure 7.7. We note that in some types the actual vertex angles are not determined uniquely, only the relations among them. As a result, the shape of the pentagons of those types can

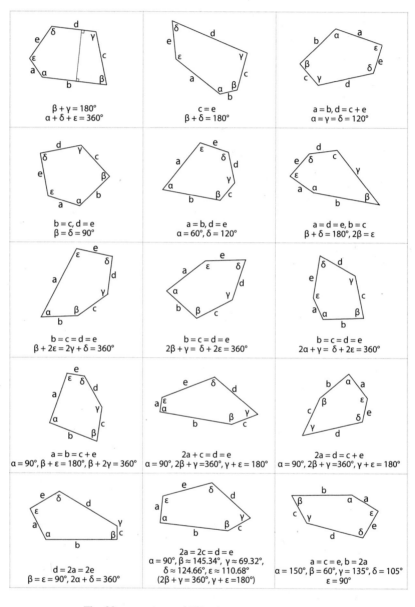

FIGURE 7.7. The fifteen pentagonal tiling types

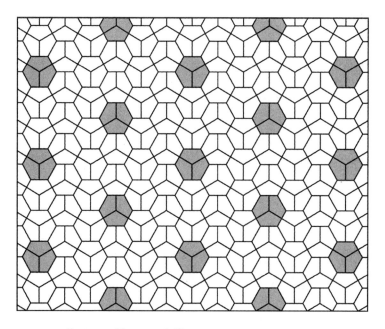

FIGURE 7.8. Pentagonal-hexagonal tiling

be varied in a continuous manner, even as the relations among the angles remain the same.[5]

★ ● ★

Somewhat surprisingly, there seems to exist a close affinity between *nonregular* pentagons and *regular* hexagons. To see this, I use a kind of reverse-engineering process. We know that congruent regular hexagons tile the plane. Now divide each hexagon into three congruent pentagons by joining the midpoints of nonadjacent sides to the hexagon's center (figure 7.8). The resulting pattern is a pentagonal tiling, although the unit cell—the smallest complete unit of the tiling—is a hexagon.

Another example of pentagonal-hexagonal tiling is the Cairo tessellation, so called because it appears frequently

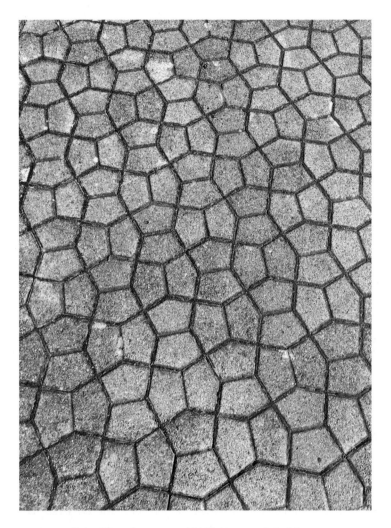

FIGURE 7.9. Cairo tiling of a pavement in Matten near Interlaken, Switzerland

in mosques across the Middle East and North Africa. Its unit cell is a nonregular hexagon with four lateral sides of equal length and two shorter bases, also of equal length; four of these nonregular pentagons fill the space of each hexagon (figure 7.9).

FIGURE 7.10. Circular tiling with congruent pentagons

★ ⬠ ★

From any of the fifteen types of pentagonal tiles we can construct tessellations that have translational symmetry and/or reflection symmetry. But if we drop these symmetry conditions, many more tessellation patterns become possible; figure 7.10 shows a beautiful pentagonal tiling with a sixfold *rotational* symmetry (note that the pentagons are still congruent but the tessellation lacks translational periodicity).

★ ⬠ ★

FIGURE 7.11. Sphere and truncated icosahedron

All the tilings we have considered so far are confined to a flat surface, a Euclidean plane. But we can also imagine spaces in which other types of tessellations are possible. The example that immediately comes to mind is the surface of a soccer ball. It has twelve regular pentagons, each surrounded edge to edge by five regular hexagons (of which there are twenty). The sphere with this tessellation is actually the projection of an inscribed solid, the *truncated icosahedron*, one of the thirteen solids attributed to Archimedes and named after him (figure 7.11).[6] And on a hyperbolic plane—the interior of a circle in which distances get ever smaller as we approach the circumference—an infinite number of regular pentagons can perfectly tile the plane (figure 7.12). It is also possible to tile a hyperbolic plane with nonregular pentagons; figure 7.13 shows an unusual pentagonal tiling with sevenfold centers of symmetry. Considered as two-dimensional non-Euclidean spaces, they allow for tilings that are impossible on the flat Euclidean plane.

★ ⬟ ★

Thus far our tilings have been made of identical, congruent pentagons. But a whole new world of tilings opens up to us if we combine the pentagons with other convex poly-

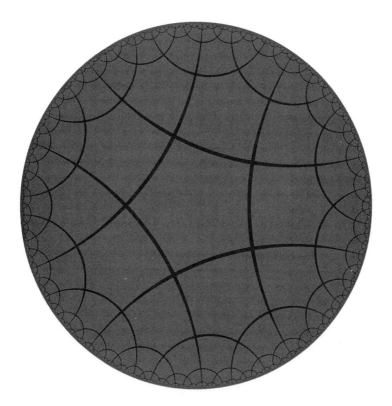

FIGURE 7.12. Pentagonal tiling on a hyperbolic plane

gons. This possibility intrigued mathematician, theoretical physicist, and cosmologist Sir Roger Penrose (b. 1931). In the 1970s he began playing with the idea of mixing pentagons with other polygons. What started as a recreational activity soon became a lasting fascination. In 1974 Penrose discovered a tiling with just two kinds of rhombi, "fat" and "lean." To his surprise, he found out that the two shapes can be combined to tile the plane in an aperiodic pattern—meaning that it cannot replicate itself by translation—and yet the entire structure possesses fivefold and tenfold symmetries about a single center.[7] A beautiful mural exhibiting this tiling can be found at the Clore Garden of Science

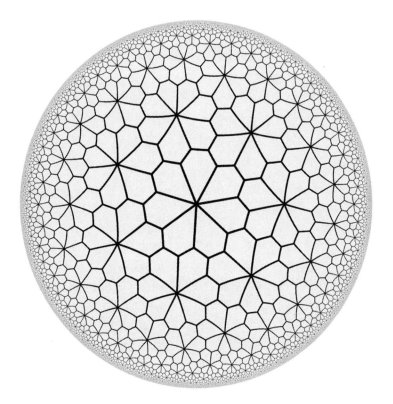

FIGURE 7.13. Hyperbolic tiling with sevenfold symmetry centers

at the Weizmann Institute of Science in Rehovot, Israel (see plate 13).

Another type of Penrose tiling uses two quadrilaterals obtained by splitting a 36-degree rhombus into two pieces, one shaped like a kite and the other like a dart. The kite has vertex angles of 72, 72, 72, and 144 degrees, whereas the dart has angles of 36, 36, 72, and 216 degrees. The kite can be split along its line of symmetry into two triangles, each with angles 72, 72, and 36 degrees—a golden triangle (see figure 7.14 and plate 14). So after our long detour from the regular pentagon, the golden ratio has finally reentered the scene.

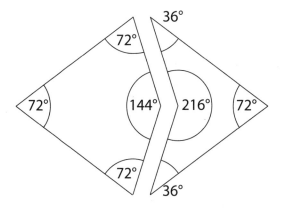

FIGURE 7.14. Kites and darts

The geometry of tessellations is perhaps the most visible link between mathematics and the arts. Long before Karl Reinhardt identified the first five of the fifteen types of pentagonal tilings, artists and artisans played with these shapes and used them to tile the facades and floors of houses of worship, particularly in the exquisite architecture of Islamic shrines. Since Islam strictly forbids the depiction of human faces, Muslim artisans found in the art of tessellation a unique way to express the infinitude of the divine. A striking example can be seen in the ancient tiles at the Bibi-Khanym mosque in Samarkand, Uzbekistan, dating back to the early fifteenth century (see plate 15). The design is a nonperiodic combination of nearly regular pentagons and nonregular hexagons (some of them concave), laced together in a dazzling display of color and shape.[8]

NOTES AND SOURCES

1. Actually, *any* triangle and *any* quadrilateral will tile the plane.
2. Regarding Reinhardt, here is an interesting aside: While involved in his research, he was also teaching mathematics at various gymnasiums (high schools). In 1934 Reinhardt published a textbook, *Methodische Einführung*

in die Höhere Mathematik (*A Methodical Introduction to Higher Mathematics*) in which he expressed his belief that students would find it easier to grasp the concept of area under a curve, rather than that of the slope of a curve. Consequently, Reinhard believed that integral calculus should be taught *before* differential calculus (regarding differentiation as the inverse of integration), rather than in the reverse order as we do today. (Source: J. J. O'Connor and E. F. Robertson, "Karl August Reinhardt," at http:// mathshistory.st-andrews.ac.uk/Biographies/Reinhardt.html.)

3. R. B. Kershner, "On Paving the Plane," *American Mathematical Monthly*, vol. 75, no. 8, October 1968, pp. 839–44; https://mth487.files.wordpress.com /2015/10/on-paving-the-plane_kershner.pdf.

4. As of this writing, Rao's computer search is still awaiting independent verification.

5. This information on the history of tiling is culled from several web articles: "Pentagonal Tiling Proof Solves Century-Old Math Problem" by Natalie Wolchover, https://www.quantamagazine.org/pentagon-tiling-proof-solves -century-old-math-problem-20170711/; "Marjorie Rice's Secret Pentagons" by Natalie Wolchover, https://www.quantamagazine.org/marjorie-rices-secret -pentagons-20170711/; and Doris Schattschneider, "Marjorie Rice (16 February 1923–2 July 2017)," *Journal of Mathematics and Art*, vol. 12, no. 1, 2018, pp. 51–54; https://www.tandfonline.com/doi/full/10.1080/17513472.2017 .1399680?src=recsys.

   You can find many more examples of pentagonal tilings—including animations that show how some tilings can generate an unlimited number of variations—in the online article "Pentagonal Tiling" at https://en.wikipedia .org/wiki/Pentagonal_tiling.

6. A fourteenth member was discovered in 2009 by Branko Grünbaum.

7. See the article "Penrose Tiling" at https://en.wikipedia.org/wiki/Penrose_tiling. As we were writing this book, it was announced that Sir Roger Penrose had been awarded the Nobel Prize in Physics for his groundbreaking work in cosmology.

8. See the article "New Light on Ancient Patterns" by Jeremy Manier, *Chicago Tribune*, February 23, 2007.

CHAPTER 8

The Discovery of Fivefold Symmetry in Crystals

> We find that crystals, which are repetitive
> assemblages of molecules, never have regular five-
> sided faces. In fact, no inanimate form exhibits
> pentagonal symmetry. No regularly pentagonal
> snowflake has ever fallen from the sky. Only
> animate forms ... have shapes with five equal sides.
>
> **—PETER S. STEVENS, *PATTERNS IN NATURE* (1974)**

IN STARK CONTRAST TO the prevalence of fivefold symmetries in the organic world—as in five-petaled flowers or
five-armed starfish—the same symmetries are completely
absent in the inorganic world of rocks and minerals. Or at
least that was the unanimous agreement among physicists,
mineralogists, and geologists for two hundred years—until
1982, when everything changed.

Unlike mathematics or astronomy, crystallography is
a relatively new science. Its founding fathers were two
Frenchmen, the priest-turned-mineralogist René-Just
Haüy (1743–1822) and the physicist Auguste Bravais
(1811–1863). Haüy (pronounced a-yoo-eé, or hah-wee)
became interested in mineralogy through an acquaintance
with a fellow priest who, as a hobby, had amassed a siz-
able collection of minerals. The two became friends, and the
priest allowed Haüy to examine a specimen of his collection, a calcite crystal. To Haüy's horror, the piece slipped

out of his hands and fell to the floor, shattering into multiple fragments.

But the accident turned out to be a blessing: while examining the broken pieces, Haüy noticed that they all could be cleaved along neat, planar faces that met at a constant angle. He examined other minerals and found the same feature: when cutting away their outer crust, what remained was a crystal with planar faces that formed specific angles with neighboring faces. Moreover, the same angles showed up in the same type of mineral regardless of where it was found or how it had grown. This led Haüy to propose that crystals are made of basic building blocks—unit cells of atoms arranged in a specific order that repeat regularly in all directions, forming a periodic three-dimensional lattice. The faces of these unit cells display certain rotational symmetries that leave the face unchanged from its original position. Haüy found that the only possible symmetries unit cells can have are two-, three-, four-, and sixfold rotational symmetries, and none other. This would be enshrined as a fundamental law of crystallography.

The next major contribution to crystallography was made by Auguste Bravais: in 1848 he showed that all crystal lattices can be classified into just fourteen classes according to the geometry of their unit cells (figure 8.1).[1] The Bravais classification exhausted the number of possible unit cells that can form a periodic lattice of atoms, a crystal. Notice that the symmetries of the Bravais lattices are exactly those that Haüy had discovered sixty years earlier. No other symmetries were present.

The Bravais classes provide us with a glimpse into the atomic structure of their respective lattices; but to discover the actual arrangement of atoms in a lattice, more advanced techniques than cleaving a crystal's faces were needed. A major breakthrough in that direction was made by the German physicist Max Theodor Felix von Laue (1879–1960):

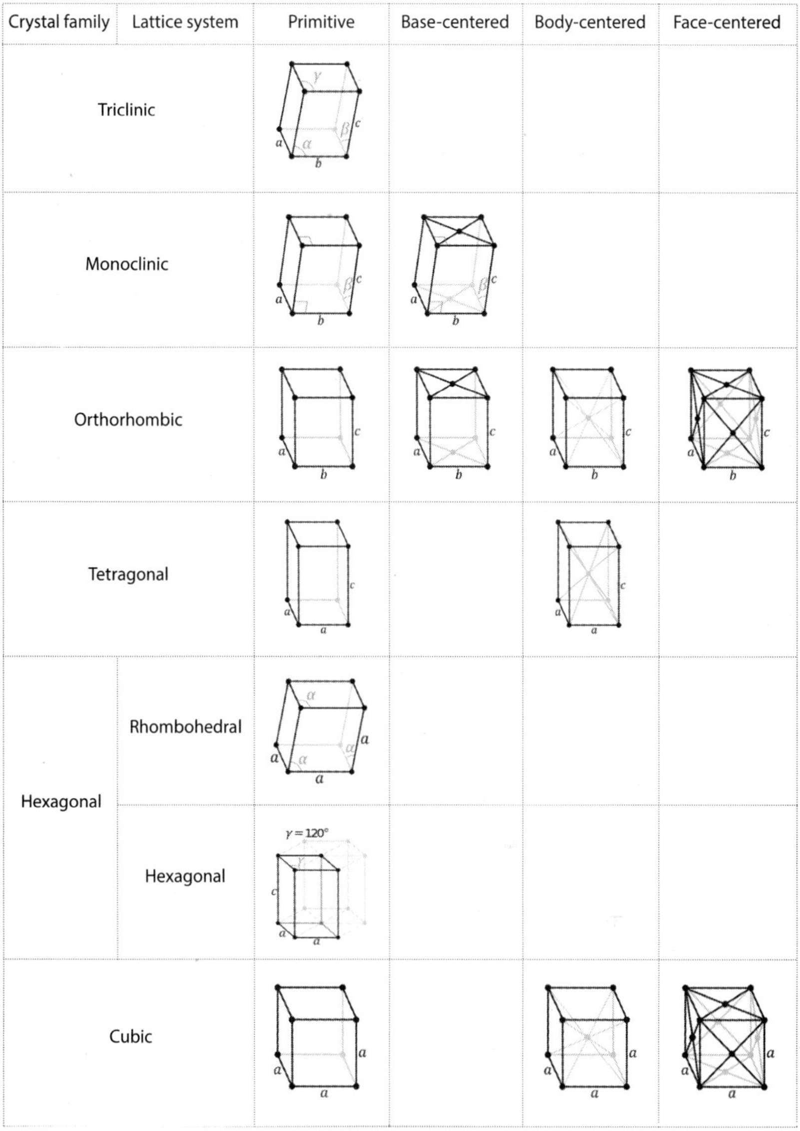

Crystal family	Lattice system	Primitive	Base-centered	Body-centered	Face-centered
Triclinic					
Monoclinic					
Orthorhombic					
Tetragonal					
Hexagonal	Rhombohedral				
	Hexagonal				
Cubic					

FIGURE 8.1. The fourteen Bravais lattices

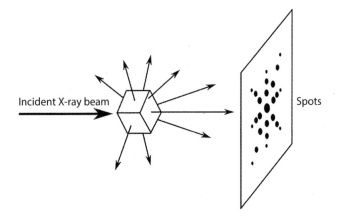

FIGURE 8.2. Diffraction pattern

in 1912 he successfully obtained the first X-ray diffraction images of a crystal. A diffraction pattern can be seen when you drop two stones into a pond: upon impact, each stone creates a system of concentric waves that spread outward independently of the other system. When the two systems meet, they interfere with one another, alternately enhancing or canceling each other. The same thing happens when a beam of light is aimed at an opaque screen with two narrow slits cut into it: the light waves pass through the slits and emerge on the other side as two independent systems of waves that interfere with one another, forming bright and dark fringes. From this diffraction pattern it is possible, in principle, to reconstruct the geometry of the two slits—their mutual separation and relative orientation in space.

To be effective, however, the wavelength of the impacting waves must be no larger than the size of the object that causes them to diffract, like the width of the two slits in the example just given. In fact, the smaller the wavelength, the sharper the diffraction pattern. Von Laue had the idea of using X-rays—discovered just a decade earlier by Wilhelm Röntgen—to reveal the atomic structure of crystals.[2] In a crystal, the many layers of atoms are arranged in parallel planes and thus form an effective diffraction grating, causing waves to bounce off adjacent layers and interfere with each other (figure 8.2). In 1912 von Laue aimed a beam of X-rays at a crystal of zinc sulfide. He recorded the resulting diffraction pattern on a photographic plate and obtained the first

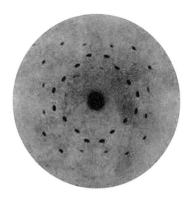

FIGURE 8.3. A von Laue diffraction image of a crystal, with fourfold symmetry

image of the atomic structure of a crystal (figure 8.3). This feat earned him the 1914 Nobel Prize in Physics.[3]

It is important to note that the dots in an X-ray diffraction pattern are not images of individual atoms, but rather locations where the interference of the diffracting waves causes them to add on to each other (constructive interference). Moreover, unlike a two-dimensional lattice, a 3-D crystal can display different symmetries, depending on how the beam is oriented with respect to the crystal. For example, each face of a cube has a fourfold rotational symmetry, clearly seen when you look down on a face along a line perpendicular to it and passing through the face's center (figure 8.4a). Likewise, each pair of diagonally opposite edges has a twofold rotational symmetry about the line connecting their midpoints (figure 8.4b). But if you look at the cube along its space diagonal (the line joining two opposite vertices), an unexpected threefold rotational symmetry reveals itself (figure 8.4c).[4] Thus, to obtain a clear picture of a crystal's symmetries, one must obtain images of many diffraction patterns, taken along different directions.

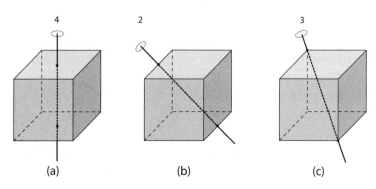

FIGURE 8.4. Symmetries of a cube

In the decades following von Laue's discovery, the internal structures of numerous crystals were obtained through X-ray and later electron diffraction (using the dual particle-wave nature of an electron beam). In all cases, only two-, three-, four-, and sixfold rotational symmetries were found, and none other—exactly as Haüy and Bravais had stated.

★ ⬟ ★

Now fast forward to 1982. Dan Shechtman (b. 1941), professor of materials science at the Technion, Israel's flagship technological institute, was spending a sabbatical at Johns Hopkins University in Baltimore, Maryland. While there, he joined a team of researchers at the US National Bureau of Standards in Washington, DC, who studied the properties of rapidly solidifying aluminum alloys as they transitioned from liquid to solid state. Using a powerful electron microscope—in which electrons, with a still shorter wavelength, replace the former X-rays—he made a startling discovery: his diffraction images of the rapidly cooled aluminum-manganese alloy Al_6Mn showed the telltale signs of tenfold rotational symmetry, which by necessity also implies a fivefold symmetry (figure 8.5). As

Shechtman recalled later, he was so puzzled by this unexpected find that he marked it with three question marks in his logbook. He checked and rechecked his findings, but the conclusion was the same: a fivefold symmetry in crystals does exist. Haüy's age-old restriction of symmetries limited to two-, three-, four-, and sixfold rotations had been proved wrong.

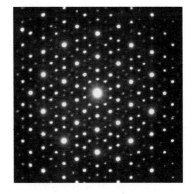

FIGURE 8.5. Shechtman's tenfold symmetry diffraction image

But the crystals that displayed this symmetry were not ordinary crystals. Existing in a state somewhere between the strictly periodic lattices of ordinary crystals and the random distribution of atoms in liquids, they were not truly periodic. They followed an orderly structure that could fill space without leaving gaps, but they did not possess translational symmetry. These crystals behave like atoms arranged in a Penrose tiling (see chapter 7).

Shechtman's discovery was sensational—and highly controversial, to put it mildly. He was repeatedly berated by colleagues who reminded him that a fivefold symmetry in crystals simply cannot exist. Among his harshest detractors was the legendary Linus Pauling, a two-time recipient of the Nobel Prize who is regarded by many as the greatest chemist of all time. Shechtman was also the subject of ridicule by his own team members, and the head of his research group went as far as asking him to leave the team. But after publishing his findings in 1984, other scientists confirmed his discovery, and in 2011 it earned him the Nobel Prize in Chemistry. To commemorate the event, the Israel Postal Authority issued a postage

FIGURE 8.6. Fivefold symmetry of an icosahedron

stamp showing a series of diminishing pentagons as an homage to Shechtman's discovery (see plate 16).[5]

★ ⬟ ★

Now, the crystal that Shechtman discovered was, in a sense, artificial—it was synthetically produced in a laboratory. Could it possibly exist also in nature? This question occupied the mind of Paul J. Steinhardt (b. 1952), then a physics professor at the University of Pennsylvania in Philadelphia and later a cosmologist at Princeton University, who, inspired by the Penrose tilings, became interested in crystallography. He teamed up with Dov Levine (b. 1958), at the time a physics doctoral student at Penn State and later professor of physics at the Technion. The two began their search by constructing styrofoam models of various 3-D objects, with the goal of forcing them to form a periodic structure. They were particularly interested in the icosahedron, the last of the five Platonic solids, with twenty faces of identical equilateral triangles (see figure 1.4). This object displays an unexpected pentagonal fivefold symmetry, which can be seen when you look down on a vertex (figure 8.6). According to the laws of Haüy and Bravais, a lattice cannot display this symmetry.

After much trial and error, Steinhardt and Levine found out that their object can indeed form a lattice, but only when combined with other objects. And even then, the lattice does not have the translational symmetry of an ordinary crystal; it follows a regular but nonperiodic order, analogous to the 2-D Penrose tilings. They called this type of lattice a quasiperiodic crystal, or *quasicrystal* for short.

Still, to be convinced that such an object could really exist, Steinhardt needed something more than a styrofoam construct. His team therefore created a computer model of the diffraction pattern that, according to their prediction, should be produced by the rapidly solidified aluminum-manganese alloy. When Shechtman's diffraction image was published in 1984, it exactly matched Steinhardt and Levine's theoretical model. The two teams, though stationed just one hundred miles apart, were unaware of each other's work, as neither team was ready to publish its findings. Yet one team made a prediction based entirely on theory, and the other, practically simultaneously, confirmed it experimentally. It was science at its best.

★　◆　★

But the story does not end here. The quasicrystal alloy that Steinhardt had predicted and Shechtman had discovered was still an artificial object—it was synthesized in the laboratory. Could quasicrystals, perhaps, be formed naturally? This was the new focal point of Steinhardt's interests. The cosmologist-turned-mineralogist now became a detective, doing an exhaustive internet search of mineral collections around the world, with the hope of finding a sample of a quasicrystal structure. Finally, his search bore fruit when Luca Bindi, then a mineralogist at the Museum of Natural History of the University of Florence in Italy, found in the institute's collection a tiny, 3 mm wide sample of the alloy $Al_{65}Cu_{20}Fe_{15}$ (the subscripts indicate the relative number of atoms of aluminum, copper, and iron, respectively). He sent it to Steinhardt, who examined it under his electron microscope and confirmed its icosahedral symmetry.

But where did this sample come from before it reached the collection in Florence? No clear record could be found of its place of origin, except for a cryptic note that seemed to trace it to a faraway corner of Siberia. After traveling to half a dozen countries and interviewing numerous "people of interest," Steinhardt finally nailed down its origin to a remote creek in the Kamchatka Peninsula in eastern Siberia, just south of the Arctic Circle. His detective work now led him to organize a geological expedition to try to find the rare mineral at the original site where it had been discovered. Fighting swarms of mosquitoes and attacks by Siberian bears, the expedition team finally located the creek and managed to excavate a few samples of the precious mineral. It was a triumphal moment, and the team members celebrated with a good drink of vodka. It was the summer of 2011, almost thirty years after Shechtman's discovery of the first known quasicrystal, and, coincidentally, the same year that he was awarded the Nobel Prize.

How had this sample been formed? Ordinarily, nature cannot replicate the rapid cooling process that created this mineral in the lab, but it could be formed as a result of an enormous impact pressure, such as when a large meteorite slams into the Earth. Immediately, the Moon's craters come to mind. Thanks to the lunar samples brought back by the Apollo astronauts, we know that the lunar craters were formed in the early days of the solar system by relentless meteoric bombardment. The same process is also believed to have happened here on Earth, but due to erosion by our atmosphere and oceans, few traces of it remain. It is now believed that the Kamchatka sample may be a rare survivor of this ancient cataclysm.

In due time, many other alloys were found with quasicrystal lattices of icosahedral symmetry. For example, in 1987 An-Pang Tsai and his team at Tohoku University in Sendai, Japan, discovered an aluminum-copper-iron quasicrystal, $Al_{65}Cu_{20}Fe_{15}$, with perfectly shaped pentagonal faces.[6] This was followed a decade later when Ian Fisher and his team at Stanford University identified a quasicrys-

FIGURE 8.7. Quasicrystal $Ho_9Mg_{34}Zn_{57}$

tal with the unusual chemical composition $Ho_9Mg_{34}Zn_{57}$ (Ho is holmium, a rare-earth element of the lanthanide series with atomic number 67; the other elements are magnesium and zinc).[7] A photograph of this alloy, showing a tiny grain with perfectly shaped pentagonal faces and edges 2.2 mm long, has become an iconic image (figure 8.7). Soon quasicrystals with other formerly forbidden symmetries were discovered, proving that what had once been considered impossible had become commonplace. Indeed, quasicrystals are already finding practical uses: a new nonstick coating material for cookware and frying pans is based on quasicrystals. Whether this novel form of matter will lead to a new industrial revolution akin to the plastic revolution of the mid-twentieth century remains to be seen.[8]

NOTES AND SOURCES

1. However, see the online article "Crystallography—Defining the Shape of Our Modern World" by Vera V. Mainz and Gregory S. Girolami (http://scs.illinois.edu/xray_exhibit/). I am quoting from this article: "Recent scientific scholarship has revealed that his [Bravais's] results were actually discovered twice before: first by Moritz Frankenheim in 1826 and again by Johann Hessell in 1830. Bravais deserves credit, however, for devising a rigorous proof of the result, and bringing it to the attention of the scientific community."
2. A beam of X-rays can behave either as a stream of particles or as a continuous wave with an extremely short wavelength of between 10^{-9} and 10^{-6} cm.
3. Von Laue was one of the very few German scientists who actively resisted the Nazi policies before and during World War II. In an act of protest, he resigned his post at the University of Berlin in 1943, a position he had held since 1919. After the war, he became director of the Max Planck Institute for Physical Chemistry in Munich.
4. I still remember my surprise when I made this "discovery" as an undergraduate physics student, as it defied my mental image of a cube. If you disregard the edges, a cube, when viewed along its space diagonal, will actually look like a hexagon.
5. In 2019 my wife and I attended a public talk by professor Shechtman at the Begin Center in Jerusalem. In the first half of his talk, he presented a simple, clear outline of the structure and symmetry of crystals. Then he described the agonizing years of ridicule he had to endure from his colleagues for his unconventional discovery. Mentioning Linus Pauling, Shechtman said, "He kept harassing me for ten years, stopping only when he died in 1994" (my own translation from the Hebrew). After his talk, we spoke with him briefly. He was pleased that two of his fellow Technion graduates attended his talk (my wife and I graduated just a few years before him).

    For more on Shechtman's discovery, see the 2011 article "Quasicrystals Scoop Prize" by Laura Howes at https://www.chemistryworld.com/features/quasicrystals-scoop-prize/3004748.article.
6. Source: "A Stable Quasicrystal in Al-Cu-Fe System" at https://iopscience.iop.org/article/10.1143/JJAP.26.L1505/meta.

    As we were preparing this book for publication, we learned of the untimely passing of Professor An-Pang Tsai at the age of 60. His obituary can be found at https://www.iucr.org/news/newsletter/volume-27/number-2/an-pang-tsai.
7. The numbers following each of the three elements do *not* represent atoms; as Ian Fisher explained to me, "Because quasicrystals don't have a unit cell, the composition is expressed as a ratio of elements summing to 100."
8. The narrative in this chapter is partially based on *The Second Kind of Impossible: The Extraordinary Quest for a New Form of Matter* by Paul J. Steinhardt, listed in the bibliography.

An Unsolved Mystery

IN 1982 AT THE TIME OF THE LAST RESTORATION of the St. Pierre Cathedral in Geneva, Switzerland, a fourth-century Roman dodecahedron-shaped metallic object was discovered (figure S4.1). Its twelve pentagonal faces are made of silver and marked with the twelve signs of the zodiac (figure S4.2). The core is made of lead, and the object weighs 297 grams. It has a width of 35 mm, and the side of each face is 15 mm long. It is known as the *Geneva dodecahedron*.

This was a remarkable finding, but what was the significance of this object? What was it used for? In the absence of any written documentation, the origin and purpose of the Geneva dodecahedron remains a mystery. Theories abound: a religious artifact, an ornamental charm, or perhaps a die serving the same purpose as it would today—to draw random numbers, in this case between one and twelve.

This was not the first discovery of an ancient dodecahedron-shaped artifact. A much more elaborate object was found in Germany in 1739, and over one hundred similar objects have since been found all over the northwestern reaches of the Roman Empire; for this reason they are known as *Gallo-Roman* dodecahedrons. They are made of copper alloy, range in size from 4 to 11 cm, and have the shape of a dodecahedron with a circular hole in each of the twelve faces; their interior is hollow (figure S4.3). The holes are of different sizes, raising the speculation

FIGURE S4.1. The Geneva dodecahedron

that the object was used as a knitting loom, or as a personal attacking or defensive device that a soldier would put on the palm of his hand. But it is also possible that it was simply a decorative object or an amulet that promised its owner good luck. Several of these artifacts have been uncovered in caches of Roman coins, so they must have been of great value to their owners. In one case, an icosahedron was discovered near the town of Arloff, Germany; it is now displayed at the Rheinisches Landesmuseum in Bonn, along with several Roman dodecahedra. How these objects were cast, and what their purpose was, remains a mystery.

But why did the artisans who crafted these objects choose a dodecahedron, rather than the much simpler cube or octahedron? The answer may be found in the special role the dodecahedron played in ancient Greek mythology. Plato (ca. 427–347 BCE), after whom the five regular solids are named, has associated the tetrahedron, hexahedron (cube), octahedron, and icosahedron with the four elements that, according to the Pythagoreans, made up the universe: fire, earth, air, and water, respectively. But what role should then be given to the dodecahedron, with its twelve pentagonal faces? The Pythagoreans had an answer: this object symbolized the entire universe—all the more so since its twelve faces could represent the twelve constellations of

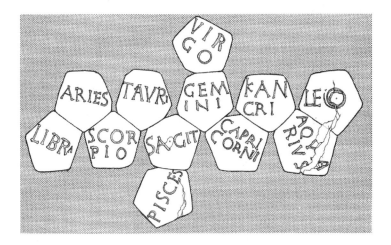

FIGURE S4.2. The twelve faces of the Geneva dodecahedron

FIGURE S4.3. A Gallic dodecahedron

the zodiac. To them it was a perfect arrangement, and they enshrined it in their mythology. The Romans, who considered themselves the bearers of the classical Greek tradition, were only too happy to adopt it in their art and craftwork.

CHAPTER 9

Oh, *That* Pentagon

The surest test of the civilization of a people…
is to be found in their architecture.

—WILLIAM HICKLING PRESCOTT,
THE CONQUEST OF PERU (1847)

AS WITH THE INORGANIC WORLD OF CRYSTALS, fivefold
symmetries are not common in architecture. One obvious
reason is that a regular pentagon cannot replicate itself
without leaving gaps—it does not tessellate the plane. Add
to this the fact that most of our home artifacts—furniture,
appliances, even cars—are basically rectangular in shape
and would not easily fit into a pentagonal room. More generally, our innate sense of orientation is rectangular, based
on left-right, forward-backward movements. A 90-degree
intersection of roads comes to us quite naturally; but a turn
of 60 degrees can be rather disorienting, as any first-time
visitor to Washington, DC, can attest to: the city's juxtapo-
sition of a rectangular and a 60-degree grid can be a challenge to navigate.

But just as with crystals, exceptions to the forbidden
fivefold symmetry can be found in architecture, and their
rarity makes them all the more interesting. As of this writing, the title for the world's tallest pentagonal building
goes to the JPMorgan Chase Tower, a seventy-five-story
skyscraper in downtown Houston, Texas. It was completed
in 1981 and reaches a height of 1,002 feet (305.4 meters).
Its original design had called for an even taller building

of eighty floors, but the Federal Aviation Administration objected for fear of endangering air traffic to and from the nearby William P. Hobby Airport. Still, even with its reduced height, it was for a while the eighth tallest building in the United States.[1]

The building's cross section, however, is not a regular pentagon. That honor goes to the Baltimore World Trade Center, located on the city's Inner Harbor. It was designed by the architectural firm Pei Cobb Freed & Partners and built from 1973 to 1977. It has thirty floors and rises to a height of 450 feet (137 meters) above the ground plaza between Pratt Street and the harbor. Unfortunately, due to its height, the structure's pentagonal shape is not easily perceived from the ground.[2]

One of the most bizarre examples of pentagonal architecture is the housing complex Ramot Polin (Hebrew for "Poland Heights") in Jerusalem, designed by Polish-born Israeli architect Zvi Hecker (b. 1931) and built in the city's Ramot neighborhood in the early 1970s (figure 9.1). Planned for 720 housing units, each in the shape of a dodecahedron, it has been described as "one of the world's strangest buildings,"[3] a neighborhood that "from afar looks like a giant beehive or a set of crystalline rocks."[4] Over the years, nonpentagonal units have been added, and tenants have made their own alterations to the complex. The Israeli architectural historian David Kroyanker describes it as "nothing but sculpture, a formal curiosity that was interesting from a geometrical point of view but lacked real content."[5] Like any controversial issue, the project triggered a fierce debate over the limits of creative architecture to address human needs.

Sadly, "human needs" have included, throughout history, countless acts of war. The need to defend a city from marauders has prompted architects and engineers to design a whole range of defensive structures—walls and fortresses,

FIGURE 9.1. Ramot Polin, Jerusalem

ditches and moats, towers and bastions—to repel an assault by an invading enemy. The shape of a fortress was often determined by the layout of the ground around it. If no topographical constraints existed, the most natural fortification was a fortified square building or a courtyard protected by a surrounding wall. Figure 9.2 shows the massive fortress known as David's Citadel, built by Ottoman ruler Suleiman the Magnificent in the years 1537–1541. It stands intact today, guarding the Jaffa Gate's western entrance to the Old City of Jerusalem.

Nevertheless, quite a few medieval fortresses were built in the shape of a pentagon. There are two explanations for the choice of this unusual shape. For one, a regular pentagon of unit side holds a greater area than a square with the same perimeter:

Perimeter of a pentagon of unit side: 5
Area of a square of perimeter 5: $(5/4)^2 \sim 1.563$
Area of a unit pentagon: ~ 1.720

FIGURE 9.2. David's Citadel, Jerusalem

(for the area of a unit pentagon, see the last equation in chapter 5). So the pentagon encloses about 10 percent more area than a square with the same perimeter.

We can find the area-to-perimeter ratios for other regular polygons:

Perimeter of a hexagon of unit side: 6
Area of a square of perimeter 6: $(6/4)^2 = 2.250$
Area of a unit hexagon $= 3\sqrt{3}/2 \sim 2.598$,

an increase of about 15.5 percent over the square.

Perimeter of an octagon of unit side: 8
Area of a square of perimeter 8: $(8/4)^2 = 4$
Area of a unit octagon $= 2(1 + \sqrt{2}) \sim 4.828$,

an increase of 20.7 percent. So it seems that as we increase the number of sides, the area relative to a square of equal perimeter steadily increases.

How far can we go? You might guess that if we could continue this process forever, we would get a "polygon" with an infinite number of equal sides—that is, a circle. The circumference of a circle of unit radius is 2π, so a square with the same perimeter would have an area of $(2\pi/4)^2 \sim 2.467$. By comparison, the area of the unit circle is $\pi \sim 3.142$, an increase of about 27.3 percent over the unit square. This indeed is the maximum possible gain in area we can achieve with this process. It is a special case of the *isoperimetric problem*, which states that, of all closed planar curves with the same perimeter, the circle has the largest area.

So why not increase the number of sides of a fortress to any arbitrary number—even to a circle—so as to increase the area we get for a given perimeter? This would result in fewer bricks needed to build the wall and thus reduce the cost of material and labor.[6] But there's a catch: a polygonal wall with a large number of sides would require a watchtower at each corner, increasing not only the cost of construction but also the number of soldiers required to defend it. A pentagonal shape would be an acceptable compromise, enclosing a slightly smaller area than a hexagonal or octagonal fortress but requiring just five defensive watchtowers.

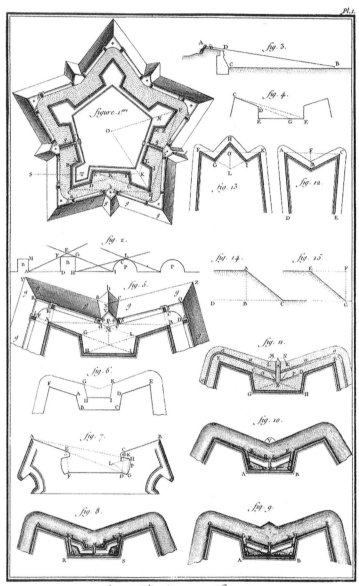

Pl. 1.

Art Militaire, Fortification.

FIGURE 9.3. A pentagonal fortress (probably from Diderot's *Encyclopedia*, Paris, 1762)

And there was one other consideration: the "dead zones" at each corner, those areas that would be inaccessible to cannon fire from the corner bastions. A large body of literature was devoted to this one question; indeed, mathematics texts of the sixteenth and seventeenth centuries often had entire chapters on "military mathematics," with detailed calculations of the range of cannon batteries, the optimal visibility from corner watchtowers, and the shape and area of dead zones where artillery fire would be ineffective. We don't need to go into the details of these calculations; suffice it to say that a pentagonal fortress was the preferred shape for many fortresses in the sixteenth and seventeenth centuries (figure 9.3).

Regrettably, only a few of these forts have survived. Notable among them is the magnificent pentagonal Citadel of Jaca, a town of some 13,000 residents in the province of Huesca in northeastern Spain and capital of the kingdom of Aragon until 1097 (see plate 17). The fortress is located at the confluence of two ancient roads, one crossing the Pyrenees to the north into France, the other leading south to the heart of Spain. Construction began in 1592 on the order of King Phillip II, whose architect, Tiburcio Spannocchi (1541–1609), "followed the canons of military architecture derived from the use of artillery."[7] It was completed in 1613 by the king's son Phillip III. The central pentagonal parade ground and the five barracks that line up the defensive walls were restored in 1968, as were the "arrow" bastions at each corner and a drawbridge over the surrounding moat. One of the barracks houses the Museum of Military Miniatures, boasting a collection of over 32,000 lead soldier figurines. The citadel also houses the headquarters of the Ski and Mountaineering Division of the Spanish Army.[8] The complex includes a Baroque-style military chapel with a stone doorway, built in the second half of the seventeenth century. The chapel is dedicated to St. Peter (hence the Spanish name Castle of San Pedro).

★ ♠ ★

We mentioned earlier Baltimore's World Trade Center, citing its pentagonal cross section. But the city has another famous pentagonal structure to boast about, dating to a much earlier time: Fort McHenry, a coastal bastion located in the city's Locust Point neighborhood. Built in 1798–1800 to defend the Baltimore Harbor against a naval attack, it fulfilled its mission fourteen years later, when the British Navy attacked the fort from Chesapeake Bay on September 13–14, 1814. The saga of those two days became enshrined in American history. Watching the battle from a nearby ship in the bay, American statesman Francis Scott Key saw the US flag, with its fifteen stars and fifteen stripes, emerge intact after two days of relentless bombardment. Deeply moved by the sight, he expressed his emotions in a poem, "Defence of Fort M'Henry"; it was later set to the tune of the song "To Anacreon in Heaven," renamed "The Star-Spangled Banner," and in 1931 became the national anthem of the United States of America. The fifteen-starred flag still flies over the fort to this day, although it is a replica of the original flag.

The fort is named after American statesman James McHenry (1753–1816), who was one of the signatories of the US Constitution and later became secretary of war under Presidents George Washington and John Adams. It was continually used by the US Armed Forces until World War I and by the Coast Guard in World War II. The fort was designated a national park in 1925 and a "national monument and historic shrine" in 1939. It encompasses a total area of 43.26 acres and is surrounded by a dry moat.

A plan of the fort, drawn by William Tell Poussin in 1819, is shown in figure 9.4.[9]

★ ● ★

Now fast-forward to the twentieth century. Without a doubt, the most famous pentagonal structure in history is *the* Pentagon in Washington, DC—or more precisely, in Arlington County, Virginia, just outside the capital's border (figure 9.5). It houses the US Department of Defense and is the world's largest office building, with a total usable area of 6,500,000 square feet (about 600,000 m²). Each of its five facades is 921 feet (280 m) long, its perimeter is 7/8 of a mile (1.4 km), and it encompasses a total area of thirty-four acres (roughly 137,000 m²), including a five-acre (~ 20,000 m²) central pentagonal court and a parking lot for 10,000 cars. Its interior comprises four rings of buildings parallel to the exterior building, each with five floors above ground (plus two underground levels) and five ring corridors per floor. The total length of its corridors is 17.5 miles (about 28 km), and yet it takes only seven minutes to walk between any two points in the building. The complex accommodates approximately 23,000 military and civilian employees and another 3,000 nondefense support personnel.[10] The building is so huge that one cannot perceive its pentagonal shape when facing any of its facades; only an aerial view can show its real shape.

The vast complex was designed by two architects, George Edwin Bergstrom (1876–1955) and Colonel David Julius Witmer (1888–1973), following President Franklin D. Roosevelt's decision to consolidate the seventeen buildings that had comprised the then Department of War (today the Department of Defense). The preliminary design and drafting took just thirty-four days. Construction began on

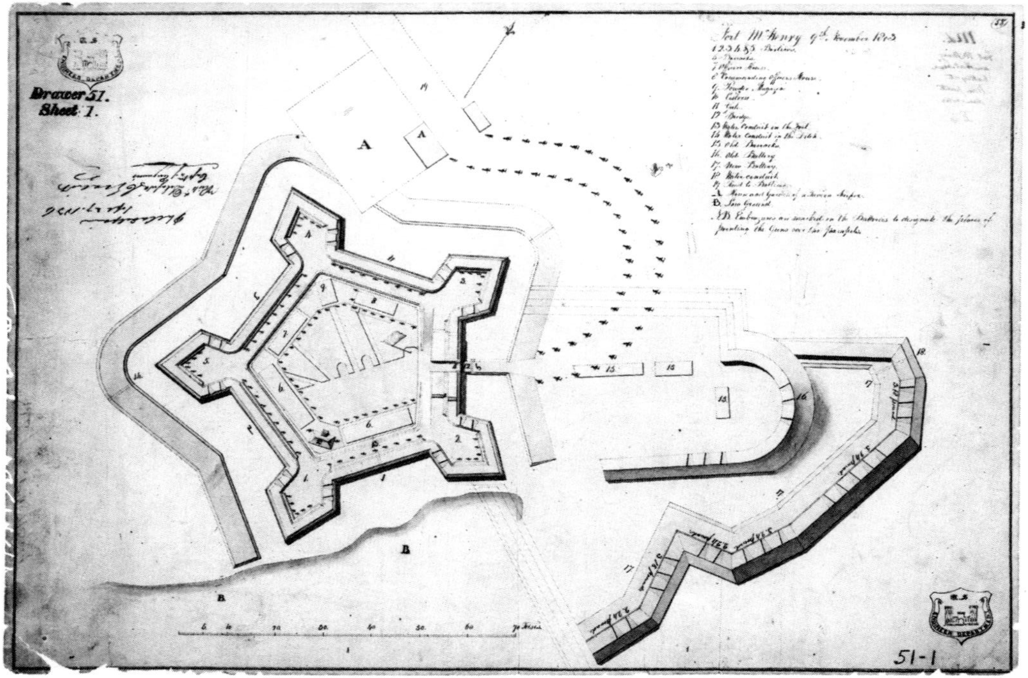

FIGURE 9.4. Plan of Fort McHenry, 1819

FIGURE 9.5. The Pentagon, Washington, DC

September 11, 1941, and was completed on January 15, 1943, at a total cost of $83,000,000 (about $1.2 billion in today's value). Overseeing the project for the US Army was Colonel Leslie Groves, who would soon be heading the Manhattan Project.

These are all impressive statistics, but why was this colossal structure built in the unusual shape of a pentagon? Theories abound: a tribute to the ancient Pythagorean emblem, an attempt to follow in the tradition of medieval pentagonal fortresses, or an effort to achieve the most efficient use of available space. The truth is much more prosaic: When the planning for the building began, the only suitable lot within reasonable distance from the capital was tucked between the Potomac River, a nearby tributary, and three access roads—an area roughly pentagonal in shape. So a pentagon it would be, prompting Hermann Weyl, in his classic book *Symmetry*, to remark, "By its size and distinctive shape, it provides an attractive landmark for bombers."[11] Fifty years after he wrote those words, on September 11, 2001, his premonition came true when American Airlines flight 77 was hijacked by terrorists and crashed into the west wing of the Pentagon, killing 189 people. It was the sixtieth anniversary of the start of construction of the building, to the day. This, too, has now become enshrined in the history of the shape we call the pentagon.

NOTES AND SOURCES

1. Source: "The Skyscraper Center: JPMorgan Chase Center" at https://www.skyscrapercenter.com/building/jpmorgan-chase-tower/472. See also the article "JPMorgan Chase Tower (Houston)" at https://en.wikipedia.org/wiki/JPMorgan_Chase_Tower_(Houston).
2. Source: http://www.waymarking.com/waymarks/WMGNPE_TALLEST_Regular_Pentagonal_Building_Baltimore_MD.
3. From Wikipedia, "Ramot Polin," April 2019, quoting http://www.travelandleisure.com/articles/worlds-strangest-buildings/5.
4. David Kroyanker, *Architecture in Jerusalem: Modern Construction Outside the Walls, 1948–1990* (Jerusalem: Keter Press, 1991; in Hebrew); quote translated by *Ha'aretz* from the 2013 article "Ad Classics: Ramot Polin/Zvi Hecker" by Gili Merin, https://www.archdaily.com/416666/ad-classics-ramot-polin-zvi-hecker.
5. Noam Dvir, "Back to the Future: A Giant Beehive Abuzz with Controversy, *Ha'aretz*, December 29, 2011; https://www.haaretz.com/1.5223121. See also https://www.youtube.com/watch?v=uIkHPKMFUuQ, which includes interviews with Hecker and Kroyanker.
6. A round wall actually surrounded the city of Baghdad when it was founded in CE 762 by Mohammad al-Mansur on the banks of the Tigris River. It had a circumference of 4 miles, with a second, concentric wall on the inside. A moat surrounded the outer wall. (Source: Violet Moller, *The Map of Knowledge: A Thousand-Year History of How Classical Ideas Were Lost and Found* [New York: Doubleday, 2019].)

7. Source: "San Pedro Castle or Citadel" at https://www.spain.info/en_US/que
-quieres/arte/monumentos/huesca/ciudadela_de_jaca.html. The words *canon*
and *cannon* are often confused for their nearly identical spelling, but in this
case the phrase "the canons of military architecture derived from the use
of artillery [i.e., cannons]" could be a play on either word. According to this
website, the Jaca fortress "is today the only fortress in the world in this style
of construction still standing."
8. Sources: "Jaca, Castle of San Pedro (St Peter) or Citadel, 16th Century" at
http://www.aspejacetania.com/lugares.php?idio=en&Id=44 and "The Citadel" at https://www.hikepyrenees.co.uk/blog/jaca-a-guide/.
9. This description is adapted from the Wikipedia article "Fort McHenry" at
https://en.wikipedia.org/wiki/Fort_McHenry.
    For other surviving pentagonal fortresses, see the following websites:
    Fort Bourtange, Groningen, the Netherlands: https://en.wikipedia.org
        /wiki/Fort_Bourtange
    Fort Belgica, Banda Neira, Indonesia: https://en.wikipedia.org/wiki
        /Fort_Belgica
    Fort Rocroi, Champagne-Ardennes, France: http://www.sites-vauban
        .org/Rocroi,746
    Indonesia's Pentagon Castles: https://www.shutterstock.com/de/image
        -photo/indonesias-pentagon-castles-1023740410
    Citadel of Pamplona, Spain: https://en.wikipedia.org/wiki/Citadel_of
        _Pamplona
    Jaffna Fort, Sri Lanka: https://en.wikipedia.org/wiki/Jaffna_Fort
    Fort Independence, Boston, Massachusetts: https://en.wikipedia.org
        /wiki/Fort_Independence_(Massachusetts)
    Fort Warren, Georges Island, Boston, Massachusetts: https://en.wikipedia
        .org/wiki/Fort_Warren_(Massachusetts)
    Fort Morgan, at the mouth of Mobile Bay, Alabama: https://en.wikipedia
        .org/wiki/Fort_Morgan_(Alabama)
    "Vatican Assassins" by Eric Jon Phelps at https://vaticanassassins.org
        /2010/06/27/the-pentagon-jesuit-military-fortress-from-spain-to
        -italy-to-the-american-empire/
    Tilbury Fort, England: https://en.wikipedia.org/wiki/Tilbury_Fort
    Fort du Mont-Valerien, Paris, France: http://www.starforts.com
        /montvalerien.html
    Kastellet, Copenhagen, Denmark: https://en.wikipedia.org/wiki
        /Kastellet,_Copenhagen
    Holt Castle, Holt, Wrexham Borough, Wales: https://en.wikipedia.org
        /wiki/Holt_Castle
    Fort of San Diego, Acapulco, Guerrero, Mexico: https://en.wikipedia
        .org/wiki/Fort_of_San_Diego
    Citadel of Turin, Italy: https://de.wikipedia.org/wiki/Zitadelle_von
        _Turin (in German)
    Citadel of Belfort, France: https://de.wikipedia.org/wiki/Zitadelle
        _Belfort (in German)

10. This description is culled from three sources: Randal Bond Truett, ed., *Washington, D.C.: A Guide to the Nation's Capital* (New York: Hastings House, 1968), p. 449; Susan Burke and Alice L. Powers, *Eyewitness Travel: Washington, D.C.* (London: DK, 2006), p. 132; and the Wikipedia article "The Pentagon" at https://en.wikipedia.org/wiki/The_Pentagon.

11. Hermann Weyl, *Symmetry* (Princeton, NJ: Princeton University Press, 1952), p. 66.

APPENDIX A

The Elementary Euclidean Constructions

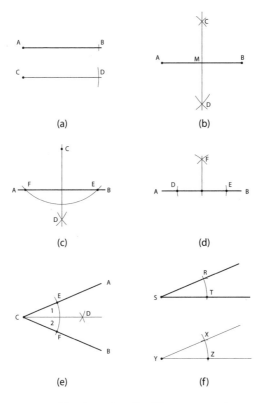

FIGURE A.1. The elementary Euclidean constructions

(a) Copying a line segment

(b) Bisecting a line segment

(c) Dropping a perpendicular to a line from a point off the line

(d) Erecting a perpendicular to a line from a point on the line

(e) Bisecting an angle

(f) Copying an angle

APPENDIX B

Three Properties of the Fibonacci Numbers

There are numerous formulas related to the Fibonacci numbers. For example,

$$F_1 + F_2 + F_3 + \cdots + F_n = F_{n+2} - 1,$$
$$F_1^2 + F_2^2 + F_3^2 + \cdots + F_n^2 = F_n \cdot F_{n+1},$$

and in particular a formula we'll shortly be using,

$$F_{n-1} \cdot F_{n+1} = F_n^2 + (-1)^n. \tag{1}$$

All these formulas can be proved by induction and the definition of the Fibonacci numbers $F_1 = F_2 = 1, F_n = F_{n-2} + F_{n-1}$ for $n = 3, 4, 5, \ldots$. We now prove three remarkable properties of these numbers:

1. The ratio of any Fibonacci number to its predecessor approaches the golden ratio as we go to higher and higher numbers; that is, $F_{n+1}/F_n \to \varphi$ as $n \to \infty$. To this end, we divide both sides of equation (1) by F_n^2:

$$\frac{F_{n-1}}{F_n} \cdot \frac{F_{n+1}}{F_n} = 1 + \frac{(-1)^n}{F_n^2}. \tag{2}$$

But $F_{n-1} = F_{n+1} - F_n$, so the left side of the last equation becomes

$$\frac{F_{n+1} - F_n}{F_n} \cdot \frac{F_{n+1}}{F_n} = \left(\frac{F_{n+1}}{F_n} - 1 \right) \cdot \frac{F_{n+1}}{F_n}.$$

At this point, let us denote the ratio F_{n+1}/F_n by x, so equation (2) becomes

$$(x-1)\cdot x = 1 + \frac{(-1)^n}{F_n^2}.$$

As $n \to \infty$, the term $(-1)^n/F_n^2$ tends to 0, so we have $(x-1)\cdot x \to 1$, or $x^2 - x - 1 \to 0$. Since the expression $x^2 - x - 1$ is a continuous function of x, its values will approach the zeros of the equation $x^2 - x - 1 = 0$, which are $(1 \pm \sqrt{5})/2$. But since we are interested here only in the positive Fibonacci numbers, we discard the negative solution and take into account only the positive solution $(1 + \sqrt{5})/2$, the golden ratio. Therefore, $x \to \varphi$ as $n \to \infty$.

As we noted in chapter 3, this result was first discovered by Simon Jacob (ca. 1510–1564) but is usually credited to Johannes Kepler, who popularized it.

2. The Fibonacci sequence begins with 1 and 1, which seems the natural way to start the progression. But nothing prevents us from starting the sequence with any two arbitrary integers a and b.[1] This, of course, will change all subsequent numbers of the sequence, but will the limiting ratio of a member to its predecessor change too? Let's see:

$$a, b, a+b, a+2b, 2a+3b, 3a+5b, 5a+8b, 8a+13b, \ldots.$$

Do the coefficients of a and b look familiar? Why, they are precisely the Fibonacci numbers 1, 1, 2, 3, 5, 8, 13, …. This shows that the original series, starting with 1 and 1, is indeed the natural choice. The generalized series can be written as

$$G_n = F_{n-2}\,a + F_{n-1}\,b. \tag{3}$$

(This formula, too, can be proved by induction. I'll leave the proof to the reader.) Now,

$$\frac{G_{n+1}}{G_n} = \frac{F_{n-1}a + F_n b}{F_{n-2}a + F_{n-1}b}.$$

Dividing the numerator and denominator by F_{n-2}, the right side becomes

$$\frac{(F_{n-1}/F_{n-2})a + (F_n/F_{n-2})b}{a + (F_{n-1}/F_{n-2})b} =$$

$$= \frac{(F_{n-1}/F_{n-2})a + (F_n/F_{n-1}) \cdot (F_{n-1}/F_{n-2})b}{a + (F_{n-1}/F_{n-2})b}.$$

As $n \to \infty$, each of the ratios in the parentheses approaches φ, so

$$\frac{G_{n+1}}{G_n} \to \frac{\varphi a + \varphi^2 b}{a + \varphi b} = \frac{\varphi(a + \varphi b)}{a + \varphi b} = \varphi,$$

showing that regardless of the two initial numbers, the ratio of a member of the sequence to its predecessor always approaches the golden ratio.

Think of it: all these algebraic transformations—a mere play with numbers and symbols—lead to a single quantity steeped entirely in geometry. This is perhaps the true justification in calling the golden ratio "divine."

3. In chapter 2 we saw how powers of φ can be converted to expressions of the form $a + b\varphi$, where a and b are Fibonacci numbers; specifically, $\varphi^n = F_{n-1} + F_n\varphi$ and $\varphi^{-n} = (-1)^n (F_{n+1} - F_n\varphi)$. It stands to reason that we should also be able to do the reverse—express any Fibonacci number in terms of powers of φ. This indeed is the case, and it leads to a remarkable formula

wrongly named after a person who was not the first to discover it—the *Binet formula*:

$$F_n = \frac{1}{\sqrt{5}}[\varphi^n - (-\varphi)^{-n}].$$

To prove it, we write again the pair of formulas

$$\varphi^n = F_{n-1} + F_n\varphi \qquad (4)$$

and

$$\varphi^{-n} = (-1)^n (F_{n+1} - F_n\varphi). \qquad (5)$$

If n is even, equation (5) becomes

$$\varphi^{-n} = F_{n+1} - F_n\varphi. \qquad (6)$$

Subtracting equation (6) from (4), we get $\varphi^n - \varphi^{-n} = F_{n-1} - F_{n+1} + 2F_n\varphi$. But $F_{n-1} - F_{n+1} = -F_n$, so $\varphi^n - \varphi^{-n} = -F_n + 2F_n\varphi = F_n(-1 + 2\varphi) = \sqrt{5}F_n$, and we get

$$F_n = \frac{1}{\sqrt{5}}(\varphi^n - \varphi^{-n}). \qquad (7)$$

But if n is odd, equation (5) becomes

$$\varphi^{-n} = -F_{n+1} + F_n\varphi. \qquad (8)$$

We now *add* equations (4) and (8), giving us $\varphi^n + \varphi^{-n} = F_{n-1} - F_{n+1} + 2F_n\varphi = -F_n + 2F_n\varphi = F_n(-1 + 2\varphi) = \sqrt{5}F_n$, so we get

$$F_n = \frac{1}{\sqrt{5}}(\varphi^n + \varphi^{-n}). \qquad (9)$$

Finally, equations (7) and (9) can be combined into a single formula,

$$F_n = \frac{1}{\sqrt{5}}[\varphi^n - (-\varphi)^{-n}], \tag{10}$$

which holds for all integer values of n—even and odd. We also note that if we change n to $-n$, equation (10) is reduced to equation (9) when n is odd, but changes its sign when n is even; that is, $F_{-n} = (-1)^{n+1} F_n$. For example, $F_{-11} = 89 = F_{11}$, while $F_{-12} = -144 = -F_{12}$.

Equation (10) is named after Jacques Philippe Marie Binet (1786–1856), a French mathematician who specialized in number theory and algebra. He found the formula in 1843, but it was already known to Abraham de Moivre, Daniel Bernoulli, and Leonhard Euler a century before. Justly or not, the name stuck.

Binet's formula is remarkable because it expresses an integer in terms of three irrational numbers, φ, $1/\varphi$, and $\sqrt{5}$. This in itself is not unusual: the sum of two irrational numbers may be rational, as the example $(1 + \sqrt{2}) + (1 - \sqrt{2}) = 2$ shows. And yet, I must confess I get a kick out of finding Fibonacci numbers with this formula. It is so strange to simply plug in the value of n and let your calculator do the rest: it always comes out right (well, at least until you reach the limit of how many digits your calculator can handle without going into scientific notation).[2] For example, $F_{12} = 144$, the number of rabbits after twelve months in Fibonacci's original statement of his famous problem (see chapter 1); after twenty-four months, the number grows to 46,368, and after thirty-six months, to 14,930,352. Fibonacci, of course, did not have the luxury of using an electronic calculator; he would have done all his calculations by hand and may have checked them on his abacus.

NOTES

1. An example is the *Lucas numbers*, named after François Édouard Anatole Lucas (1842–1891): 2, 1, 3, 4, 7, 11, 18, 29,…. Like the Fibonacci numbers, the Lucas sequence can be extended to negative numbers by using the formula $L_{n-2}=L_n-L_{n-1}$:…, −29, 18, −11, 7, −4, 3, −1, 2, 1, 3, 4, 7, 11, 18, 29,…

2. When you do this, it helps to first compute the decimal value of φ and store it in the calculator's memory, then retrieve it to complete the calculation.

A Proof That There Exist Only Five Platonic Solids

THE LAST BOOK OF *THE ELEMENTS*, book XIII, deals with the five Platonic solids. Propositions 7–11 of book XIII address various properties of a regular pentagon inscribed in a circle, and propositions 13–17 discuss the construction of the five solids. Then comes proposition 18—the very last proposition in *The Elements*. Here Euclid compares the sides of the five solids, all inscribed in a given sphere. At the end of the proposition, he says, "No other figure [polyhedron], besides the said five figures, can be constructed which is contained by equilateral and equiangular figures [i.e., regular polygons] equal to one another." That is, the tetrahedron, cube, octahedron, dodecahedron, and icosahedron are the only possible regular polyhedra. His proof is one of the longest in *The Elements*, taking up nearly six pages, and it can easily wear down a modern reader. We give here a modern proof, based on Euler's formula

$$V - E + F = 2, \tag{1}$$

relating the number of vertices V, the number of edges E, and the number of faces F of a simply connected polyhedron (a polyhedron having no holes). In a *regular* polyhedron, all faces are identical regular polygons, each having n sides. For example, the cube has six identical square faces, each with four sides, so $F = 6$ and $n = 4$; it also has eight vertices and twelve edges, so $V = 8$ and $E = 12$. Putting this into equation (1), we get $8 - 12 + 6 = 2$, as the formula requires.

For any regular polyhedron, counting the total number of edges, we have

$$nF = 2E \qquad (2)$$

because each edge belongs to two adjacent faces and is therefore counted twice. Suppose further that r edges join at each vertex (for the cube, $V = 8$ and $r = 3$). Counting again the number of edges, we have

$$rV = 2E \qquad (3)$$

because each edge connects two vertices. Substituting F and V from equations (2) and (3) into equation (1), we have

$$\frac{2E}{r} - E + \frac{2E}{n} = 2$$

or

$$\frac{1}{n} + \frac{1}{r} = \frac{1}{E} + \frac{1}{2}. \qquad (4)$$

Now we know that $n \geq 3$ and $r \geq 3$ because a polygon must have at least three sides, and at least three edges must meet at each vertex of a polyhedron. And, of course, E must be positive, so $1/E > 0$, so equation (4) becomes an *inequality*,

$$\frac{1}{n} + \frac{1}{r} > \frac{1}{2}. \qquad (5)$$

All that remains for us to do is to list the cases that this inequality allows, and rule out the rest:

Case 1: $n = 3, r = 3$: $\dfrac{1}{3} + \dfrac{1}{3} = \dfrac{2}{3} > \dfrac{1}{2}$, giving us $E = 6$: the
tetrahedron.

Case 2: $n = 3, r = 4$: $\dfrac{1}{3} + \dfrac{1}{4} = \dfrac{7}{12} > \dfrac{1}{2} \Rightarrow E = 12$: the *octahedron*.

Case 3: $n = 3, r = 5$: $\dfrac{1}{3} + \dfrac{1}{5} = \dfrac{8}{15} > \dfrac{1}{2} \Rightarrow E = 30$: the *icosahedron*.

Case 4: $n = 4, r = 3$: Since n and r play a symmetric role in equation (4), we can use case 2 and get again $E = 12$. But with the roles of n and r reversed, we get the *hexahedron*, commonly known as the cube.

Case 5: $n = 5, r = 3$: This again gives us $E = 30$, but with the roles of n and r reversed, we get the *dodecahedron*.

This exhausts all possible cases, because any higher values of n and r make the left side of inequality (5) less than 1/2 and therefore do not result in a regular polyhedron (note that for $n = r = 4$, $\dfrac{1}{E} = 0$ and E is undefined).

This symmetry between the roles of r and n in equation (4)—as also between V and F in equation (1)—means that we can interchange the roles of these variables without affecting the outcome. For example, the cube has six faces ($F = 6$) and eight vertices ($V = 8$), but if we interchange these values we get the octahedron, with eight faces ($F = 8$) and six vertices ($V = 6$). Two regular polyhedra for which the values of V and F are interchanged are called *dual*. A dual of a given polyhedron can be obtained by connecting the centers of its faces with straight lines. Thus, the cube and octahedron form a dual pair, as do the dodecahedron and icosahedron (figure C.1). The tetrahedron ($F = V = 4$) is its own dual.

★ ⬟ ★

Just as new types of planar tessellations are possible if we relax the condition that the tiles should be identical

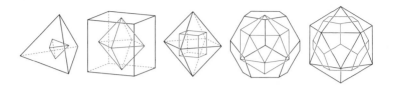

FIGURE C.1. Dual polyhedra

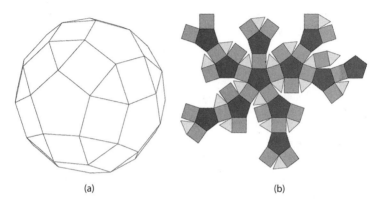

(a)   (b)

FIGURE C.2. Rhombicosidodecahedron: (a) isometric view and
(b) flattened view

regular polygons, there are likewise polyhedra with faces
of two or more types of regular polygons. For example, a
polyhedron with the impossibly long name *rhombicosidodecahedron* has twelve regular pentagonal faces, thirty
squares, and twenty equilateral triangles, making a total
of sixty-two faces (figure C.2). It is one of the thirteen *semi-
regular* or *Archimedean solids*.[1] A variant of this object, in
which the squares are replaced by rectangles and all faces
are hollow, can be constructed with plastic struts and balls;
it has become popular under the name *zome* or *zome ball*.[2]

The number of polyhedra increases even further if we
drop the requirement that they should be convex. For
instance, if the edges of each face of a dodecahedron are
extended in pairs until each pair meets at a point, we get

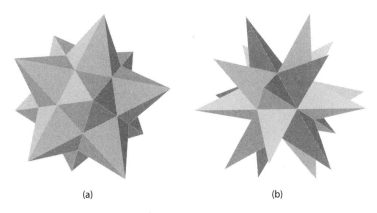

(a)   (b)

FIGURE C.3. (a) Small stellated dodecahedron. (b) Great stellated dodecahedron

the *small stellated dodecahedron*. Similarly, extending the *faces* in pairs until they meet at a line gives us the *great stellated dodecahedron*. These two solids were discovered by Johannes Kepler and are sometimes named after him. The two objects, each having the same symmetries as the icosahedron, are shown in figure C.3.

The study of these polyhedra, with their hosts of symmetries, is a fascinating subject. I refer the reader to two books: *Platonic and Archimedean Solids*, written and beautifully illustrated by Daud Sutton, and *Mathematical Models* by H. M. Cundy and A. P. Rollett; both are listed in the bibliography.

NOTES AND SOURCES

1. See, however, chapter 7, note 6. See also the article "Archimedean Solids" at https://en.wikipedia.org/wiki/Archimedean_solid.
2. See the July 2007 article "The Mathematics of Zome" by Tom Davis at http://www.geometer.org/mathcircles.

APPENDIX D

Summary of Formulas

In the following, n denotes a positive integer.

Fibonacci numbers:

$$F_1 = F_2 = 1, F_3 = 2, F_4 = 3, F_5 = 5, F_6 = 8, \dots,$$

and in general

$$F_{n+2} = F_{n+1} + F_n. \tag{1}$$

Powers of φ:

$$\varphi^2 = 1 + \varphi, \ \varphi^3 = 1 + 2\varphi, \ \varphi^4 = 2 + 3\varphi, \ \varphi^5 = 3 + 5\varphi, \ \varphi^6 = 5 + 8\varphi, \dots,$$

and in general

$$\varphi^n = F_{n-1} + F_n \varphi. \tag{2}$$

Negative powers of φ:

$$\varphi^{-1} = -1 + \varphi, \ \varphi^{-2} = 2 - \varphi, \ \varphi^{-3} = -3 + 2\varphi, \ \varphi^{-4} = 5 - 3\varphi,$$
$$\varphi^{-5} = -8 + 5\varphi, \dots,$$

and in general

$$\varphi^{-n} = (-1)^n \ (F_{n+1} - F_n \varphi). \tag{3}$$

Binet's formula:

$$F_n = \frac{1}{\sqrt{5}} [\varphi^n - (-\varphi)^{-n}].$$

Formulas related to a regular pentagon:

Pentagon of unit side:

Perimeter: 5

Length of diagonal: $\varphi = \dfrac{1+\sqrt{5}}{2}$

Area: $\dfrac{\sqrt{5(3+4\varphi)}}{4} = \dfrac{\sqrt{5(5+2\sqrt{5})}}{4} = 5\cot 36°/4$

Perimeter of inscribed pentagram: 5φ

Side of pentagon enclosed by pentagram:

$1/\varphi^2 = \dfrac{3-\sqrt{5}}{2}$

Perimeter of inscribed pentastar: $10/\varphi = 5(-1+\sqrt{5})$

Area of inscribed pentastar: $\dfrac{5}{2\sqrt{3+4\varphi}} = \dfrac{\sqrt{5(5-2\sqrt{5})}}{2}$

Pentagon inscribed in unit circle:

Length of side: $\sqrt{3-\varphi} = \sqrt{\dfrac{5-\sqrt{5}}{2}} = 2\sin 36°$

Area: $\dfrac{5}{4}\sqrt{2+\varphi} = \dfrac{5}{4}\sqrt{\dfrac{5+\sqrt{5}}{2}} = \dfrac{5}{2}\sin 72°$

Pentagon circumscribing unit circle:

Length of side: $2\sqrt{7-4\varphi} = 2\sqrt{5-2\sqrt{5}} = 2\tan 36°$

Area: $5\sqrt{7-4\varphi} = 5\sqrt{5-2\sqrt{5}} = 5\tan 36°$

Pentagonal numbers:

General formula: $P_n = n(3n-1)/2 = 1, 5, 12, 22, 35, 51, 70, 92,...$

Recursive formula: $P_n = P_{n-1} + 3n - 2$

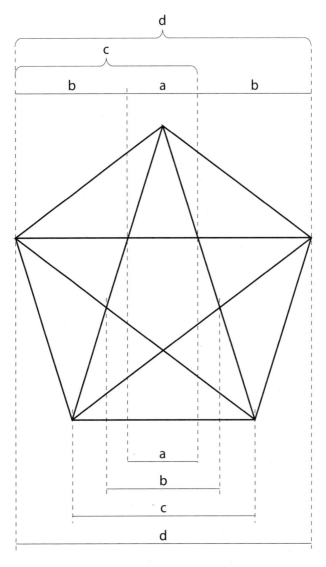

$$d : c = c : b = b : a = 1.618... = \varphi$$

FIGURE D.1. Internal relations of the pentagon/pentagram system

Solutions to Puzzles

The Shipman's Puzzle. (I am quoting from *The Canterbury Puzzles*, p. 173.)

> There are just two hundred and sixty-four different ways in which the ship *Magdalen* might have made her ten annual voyages without ever going over the same course twice in a year. Every year she must necessarily end her tenth voyage at the island from which she first set out.

The Ploughman's Puzzle. (I am again quoting from *The Canterbury Puzzles*, pp. 175–176.)

> The illustration [figure E.1] shows how the sixteen trees might have been planted so as to form as many as fifteen straight rows with four trees in every row. This is in excess of what was for a long time believed to be the maximum number of rows possible; and though with our present knowledge I cannot rigorously demonstrate that fifteen rows cannot be beaten, I have a strong "pious opinion" that it is the highest number of rows obtainable.

As an aside, we note that if we remove the center point of the inner pentagram in figure E.1, we get the solution to a much simpler puzzle: how to plant ten trees in five rows, with four trees per row.

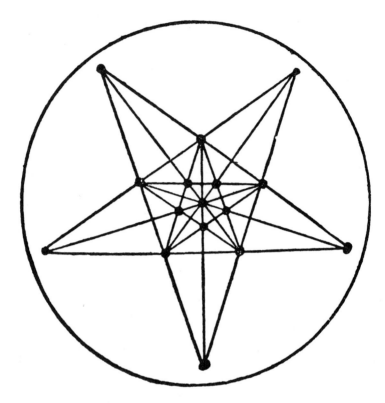

FIGURE E.1. The ploughman's puzzle: sixteen trees, fifteen rows, four trees per row

The Pentagon and Square. Here is the solution in Dudeney's own words (edited for brevity):[1]

> The pentagon is *ABCDE* [figure E.2]. By the cuts *AC* and *FM* (*F* being the midpoint between *A* and *C*, and $\overline{AM} = \overline{AF}$), we get two pieces that may be placed in position at *GHEA* and form the parallelogram *GHDC*. We then find the mean proportional between the length *HD* and the *height* of the parallelogram. This distance we mark off from *C* at *K*, then draw *CK*, and from *G* drop the line *GL*, perpendicular to *KC*. The rest is easy and rather obvious. It will be seen that the six pieces will form either the pentagon or the square.

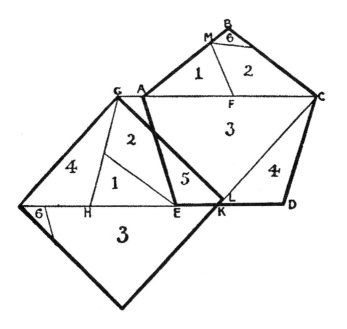

FIGURE E.2. The pentagon and square

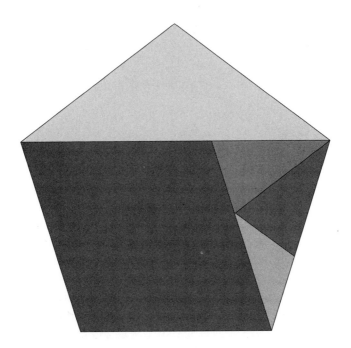

FIGURE E.3. The unscrambled pentagonal tangram

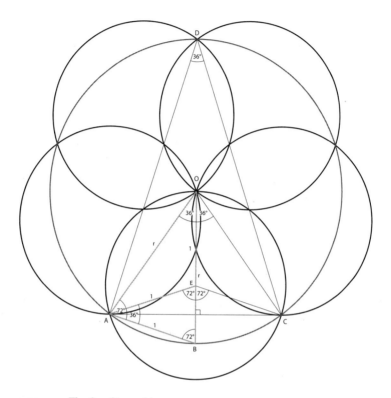

FIGURE E.4. The five-disc problem

Dudeney goes on:

> I have received what purported to be a solution in five pieces, but the method was based on the rather subtle fallacy that half the diagonal plus half the side of a pentagon equals the side of a square of the same area. I say subtle, because it is an extremely close approximation that will deceive the eye, and is quite difficult to prove inexact. I am not aware that attention has before been drawn to this curious approximation.
>
> Another correspondent made the side of his square 1¼ of the side of the pentagon. As a matter of fact, the ratio

is irrational. I calculated that if the side of the pentagon is 1—inch, foot, or anything else—the side of the square of equal area is 1.3117 nearly, or say roughly $1\frac{3}{10}$. So we can only hope to solve the puzzle by geometrical methods.[2]

How many different triangles? 35.[3]

A scrambled pentagonal tangram. See figure E.3.

The five-disc problem. $OA = (1 + \sqrt{5})/2 = \varphi$, the golden ratio. For an outline of the solution, see figure E.4. $r = OA$.

NOTES AND SOURCES

1. H. E. Dudeney, *Amusements in Mathematics*, pp. 172–73.
2. For a nice animation of this problem, see "Regular Pentagon to a Square Dissection" by Steve Phelps at https://www.geogebra.org/m/w9pXCM8E.
3. *The Moscow Puzzles* (see sidebar 3, note 3), p. 186.

BIBLIOGRAPHY

Adachi, Fumie, trans. *Japanese Design Motifs: 4,260 Illustrations of Japanese Crests*. Compiled by Matsuya Piece-Goods Store. New York: Dover, 1972.

Carroll, James. *House of War: The Pentagon and the Disastrous Rise of American Power*. Boston: Houghton Mifflin, 2006.

Conway, John H. and Richard K. Guy. *The Book of Numbers*. New York: Copernicus, 1996.

Cundy, H. Martyn, and A. P. Rollett. *Mathematical Models*. 2nd ed. London: Oxford University Press, 1961.

Edwards, Edward B. *Pattern and Design with Dynamic Symmetry*. New York: Dover, 1967.

Euclid. *The Elements*. Annapolis, MD: St. John's College Press, 1947. Translated with introduction and commentary by Sir Thomas Heath.

Gamwell, Lynn. *Mathematics +Art: A Cultural History*. Princeton, NJ: Princeton University Press, 2016.

Gardner, Martin. *Penrose Tiles to Trapdoor Ciphers: Essays on Recreational Mathematics*. New York: W. H. Freeman, 1988.

———. *Time Travel and Other Mathematical Bewilderments*. New York: W. H. Freeman, 1988.

Ghyka, Matila. *The Geometry of Art and Life*. New York: Dover, 1977.

Grünbaum, Branko, and G. C. Shephard. *Tilings and Patterns*. New York: W. H. Freeman, 1987.

Heilbron, J. L. *Geometry Civilized: History, Culture, and Technique*. Oxford: Clarendon Press, 1998.

Herz-Fischler, Roger. *A Mathematical History of Division in Extreme and Mean Ratio*. Waterloo, Ontario: Wilfrid Laurier University Press, 1987.

Huntley, H. E. *The Divine Proportion: A Study in Mathematical Beauty*. New York: Dover, 1970.

Isaacson, Walter. *Leonardo da Vinci*. New York: Simon & Schuster, 2017.

Kepes, Gyorgy, ed. *Module, Symmetry, Proportion*. London: Studio Vista, 1966.

Livio, Mario. *The Golden Ratio: The Story of Phi, the World's Most Astonishing Number*. New York: Broadway Books, 2002.

Lundy, Miranda. *Sacred Geometry*. New York: Walker, 1998.

Maor, Eli. *The Pythagorean Theorem: A 4,000-Year History*. Princeton, NJ: Princeton University Press, 2019.

———. *Trigonometric Delights*. Princeton, NJ: Princeton University Press, 2020.

Pedoe, Dan. *Geometry and the Liberal Arts*. New York: St. Martin's Press, 1976.

Posamentier, Alfred S., and Ingmar Lehmann. *The (Fabulous) Fibonacci Numbers*. Amherst, NY: Prometheus Books, 2007.

Richeson, David S. *Tales of Impossibility: The 2000-Year Quest to Solve the Mathematical Problems of Antiquity*. Princeton, NJ: Princeton University Press, 2019.

Schattschneider, Doris. *M. C. Escher: Visions of Symmetry*. New York: W. H. Freeman, 1990.

Seherr-Thoss, Sonia P., and Hans C. Seherr-Thoss. *Design and Color in Islamic Architecture*. Washington, DC: Smithsonian Institution Press, 1968.

Seymour, Dale, and Reuben Schadler. *Creative Constructions*. Oak Lawn, IL: Ideal School Supply, 1994.

Seymour, Dale, Linda Silvey, and Joyce Snider. *Line Designs*. Palo Alto, CA: Creative, 1974.

Steinhardt, Paul J. *The Second Kind of Impossible: The Extraordinary Quest for a New Form of Matter*. New York: Simon & Schuster, 2019.

Stevens, Peter S. *Patterns in Nature*. Boston: Little, Brown, 1974.

Sutton, Daud. *Platonic & Archimedean Solids*. New York: Walker, 2002.

———. *Islamic Design: A Genius for Geometry*. New York: Walker, 2007.

van der Waerden, Baertel Leendert. *Science Awakening: Egyptian, Babylonian and Greek Mathematics*. New York: John Wiley, 1963. Translated by Arnold Dresden.

CREDITS

FIGURES

FIGURE 9.1.   Ramot Polin (Poland Heights), designed by architect Zvi Hecker. Photograph by Michael Jacobson. All rights reserved. https://he.wikipedia.org/wiki/צבי_הקר#/media/קובץ:Ramotpolin01.JPG

FIGURE 9.2.   Photograph by Eli Maor

FIGURE 9.3.   From Diderot's *Encyclopedia*, Paris, 1762

FIGURE 9.4.   https://www.loc.gov/resource/hhh.md0905.photos/?sp=14

FIGURE 9.5.   USGS *The National Map*, via MSR Maps. US Geological Survey. Photograph by Mark Bisanz. https://commons.wikimedia.org/wiki/File:Pentagon-USGS-highres.jpg

FIGURE S1.1.   Design and photograph by Philip Poissant for the eipiphiny Society

FIGURE S2.1.   Eugen Jost

FIGURE S3.1.   From *The Canterbury Puzzles* by Ernest Dudeney, Dover, 1958. http://www.gutenberg.org

FIGURE S3.2.   From *The Canterbury Puzzles* by Ernest Dudeney, Dover, 1958. http://www.gutenberg.org

FIGURE S3.3.   Eugen Jost

FIGURE S3.5.   Eugen Jost

FIGURE S3.6.   Courtesy of Peter Raedschelders

FIGURE S3.7.   Eugen Jost

FIGURE S4.1.   Photograph by Daniel Berti, service cantonal d'archéologie, Geneva, Switzerland

FIGURE S4.2.   Design by Françoise Plojoux, service cantonal d'archéologie, Geneva, Switzerland

FIGURE S4.3.   AVENTICVM—Site et Musée romains d'Avenches, Switzerland

FIGURE A.1.   Eugen Jost

FIGURE C.1.   Eugen Jost

FIGURE C.2.   Eugen Jost

FIGURE C.3.   Eugen Jost

FIGURE D.1.   Eugen Jost

FIGURE E.1.   From *The Canterbury Puzzles* by Ernest Dudeney, Dover, 1958. http://www.gutenberg.org

FIGURE E.2.   From *The Canterbury Puzzles* by Ernest Dudeney, Dover, 1958. http://www.gutenberg.org

FIGURE E.3.   Eugen Jost

FIGURE E.4.   Eugen Jost

PLATES

PLATE 1.   Photograph by Eli Maor

PLATE 2.   Photograph by Eli Maor

PLATE 3.   Photograph by Eugen Jost

PLATE 4.   Eugen Jost

PLATE 5.   The Metropolitan Museum of Art, New York City, Alfred Stieglitz Collection, 1949

INDEX